Beyond the Matrix

ALSO by PATRICIA CORI

The Sirian Revelations

Vol. 1: *The Cosmos of Soul: A Wake-Up Call for Humanity*

Vol. 2: *Atlantis Rising: The Struggle of Darkness and Light*

Vol. 3: *No More Secrets, No More Lies: A Handbook
to Starseed Awakening*

The Starseed Dialogues: Soul Searching the Universe

The Starseed Awakening (Audio CD)

*Where Pharaohs Dwell: One Mystic's Journey
Through the Gates of Immortality*

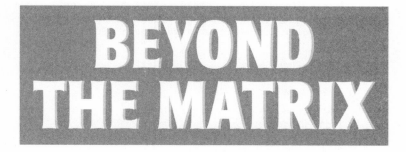

BEYOND THE MATRIX

DARING CONVERSATIONS WITH THE BRILLIANT MINDS OF OUR TIMES

Patricia Cori

North Atlantic Books
Berkeley, California

Published by
North Atlantic Books
P.O. Box 12327
Berkeley, California 94712

Cover art and design by Dragon Design UK
Tunnel illustration © shutterstock.com/Bruce Rolff;
Eagle illustration © shutterstock.com/Steven Bourelle
Book design by Brad Greene
Printed in the United States of America

Beyond the Matrix: Daring Conversations with the Brilliant Minds of Our Times is sponsored by the Society for the Study of Native Arts and Sciences, a nonprofit educational corporation whose goals are to develop an educational and cross-cultural perspective linking various scientific, social, and artistic fields; to nurture a holistic view of arts, sciences, humanities, and healing; and to publish and distribute literature on the relationship of mind, body, and nature.

North Atlantic Books' publications are available through most bookstores. For further information, visit our Web site at www.northatlanticbooks.com or call 800-733-3000.

MEDICAL DISCLAIMER: The following information is intended for general information purposes only. Individuals should always see their health care provider before administering any suggestions made in this book. Any application of the material set forth in the following pages is at the reader's discretion and is his or her sole responsibility.

Library of Congress Cataloging-in-Publication Data

Cori, Patricia.
    Beyond the matrix : daring conversations with the brilliant minds of our times / Patricia Cori.
        p. cm.
    Summary: "A discussion of diverse New Age and New Thought topics that point to the great shift in human consciousness that many say must occur for humanity and the planet to survive our current global environmental crisis"—Provided by publisher.
    ISBN 978-1-55643-893-6
    1. Consciousness. 2. Global environmental change. 3. Human beings—Forecasting. I. Title.
    BF311.C687 2010
    299'.93—dc22

                                                        2010001956

2 3 4 5 6 7 8 9 UNITED 15 14 13 12 11 10

To those who have the vision
to think outside the box,
Beyond the Matrix,
and the courage to speak their truth . . .

The Matrix is opaque and dissonant, like neon in fog.
Cluttered with the filler materials of our lives, it weighs us down in
    the density, clipping our wings, holding us captive in its web.
The Matrix is limited and confining, painting us into its grids of
    sameness: masked bandits in a hall of mirrors.
Loud and cacophonous, the Matrix tears at our inner and outer
    worlds with the noise of chaotic frenzy—
    banging the drum, sounding the horn.
It is the superfluous impulse of demand and desire,
    the prison of those who cannot see the light.
Dark and determined, the Matrix exists to enclose our minds
    in a cage, in a cube, in a box, in a corner—
    crouched in cells of our own making, where fear
    is the gatekeeper and trust is but a memory.

Beyond the Matrix is the Light.
It is clear; it is pure; it is the unchained melody.
It is the breeze upon which souls take flight.
It is the day, the eternal warmth of the sun.
It is the wonder of a cool, moonlit night.
It is a place of truth: the search, the Wisdom;
    a cosmic dance; the magic kingdom.
Beyond the Matrix lies the hope
    and the dream;
    it is the dream that keeps us alive.
It is the dream. The dream. The dream.

Contents

Renaissance Resonance

■ ■ ■ ■ ■

Patricia Cori

I have often wondered what it must have been like, in the time of the great Renaissance. Humanity, dispirited and exhausted from centuries of violence, disease, famine, and suffering, broke through the chrysalis and began to take wing. Slowly at first, signs of a new consciousness—a vision of the potential of humankind—emerged from among the privileged classes, where people could dream and dare (within the boundaries dictated by religious and royal authorities) to express their search for the nobility in all human experience.

Soon, societies at many points of the world began to experience change— a rapid shift, without precedents. The influence of art, literature, and the expanding possibilities of the new sciences extended its reach across the oceans, into the townships and cities—and a new vision brought light streaming back to a world of darkness: the Dark Ages. It breathed life back into the studies of philosophy, ancient civilizations, sacred mysteries—all fields of exploration upon which the minds of the awakening feed and are nourished.

Imagine . . .

We, the generations who walk Planet Earth in this, the beginning of the twenty-first century, are the privileged: not because we belong to the wealthy or noble classes, but because we are here. Right now. We, too, are experiencing a Renaissance—many octaves higher—as we prepare to become fully aware residents of what we are learning is an infinite multiverse, filled with countless species and unlimited populations of sentient, intelligent beings.

*We, too, are emerging from a time of relative darkness and destruction in our societies. For all our technology and achievements, the human condition, by and large, is still weighed down in suffering and discontent. We still have the privileged, the absolute minority on this vast Earth, and the struggling majority. Violence rages out of control at every corner, as that "animal" within human beings continues to seek power through greed, feeding its blood lust. The one unique difference between us and the predatory animals of our world is that **we** still kill for the sheer pleasure of the kill. And yet, we are supposedly the ones with the great intellect . . . the highest on the chain of evolution on our planet.*

In the Dark Ages of our contemporary world, diseases that we never even imagined rage, from continent to continent, forever threatening our very existence. And people are starving—that's right—starving . . . in the years we believe reflect civilization at its technological apex.

At first glance, it certainly doesn't appear that we are entering an Age of Enlightenment!

Many look out upon fields of war and see nothing but futility and hopelessness. Others gaze into the glare of the controlled media, resigning themselves to the failure of our species, believing there is simply no hope for humanity or the planet itself. The doomsday soothsayers are having a proverbial field day. They believe we are reliving Sodom and Gomorrah, and that the prophecies, which speak to the ultimate destruction of the very seed of life, are upon us: the final solution—Armageddon.

And yet . . .

Despite the mindless corruption of our modern-day figures of authority (whether "elected" or imposed on us), despite the hackneyed dictates of religions and cultures, despite the Earth Changes themselves, there is a miracle taking place, right in front of our eyes . . . just as it is being birthed within us—within every cell of our beings.

There is a rebirth of that far-reaching vision of our ancestors. The battle between logic and intuition is beginning to evolve into a new paradigm— a new science—merging with spirit. As we neutralize the duality of our very existence (as living beings in the material realm of physical space), refining the vibratory signatures of our own organisms and their imprint upon the fields of the living, we are finally recognizing how all things exist as aspects of each other. All is One in the Cosmos. All is divinely, exquisitely, harmonious when we take ourselves to the frequencies of the Higher Self. We, conscious units of that totality, are free to determine how we will play our music in the symphony of life—at every turn.

As science and spirit merge, the lines of separation between logic and intuition are blurring and all but falling away. We are learning that we have known this all along, from the great civilizations of Earth's unwritten history, down through the ages. The quintessential meaning of life has lain right before our eyes, throughout the many cycles of our existence. At times, in the cyclical rise and fall of civilizations, we have held it in our hands, precious wisdom, like a freshly cut rose; at others, we have abandoned it, in pursuit of the unreachable and vacuous ambition for personal achievement and gratification, as illusive as the clouds. We are learning, with great humility, that no matter how hard we try to render the idea of Creation complex and ultimately inconceivable, the inevitable question of existence, from the macrocosm to the microcosm, is so utterly simple. The Wisdom is coded into every level of life—every blade of grass, flower, mountain, animal, cloud . . . human; every note, melody, song, symphony.

Whatever we perceive God to be, we can surely agree that He, or She, or It has a most exquisite capacity to create the most eloquent displays of matter, organized in mathematically superlative formulas—to produce the endless array of interactive living beings on our planet and (we do believe) in worlds soon to be discovered.

Strip away method, technique, tools, and training from all spiritual disciplines and you arrive at one simple and unwavering truth: everything that exists is energy—pure energy. Everything is consciousness; everything is vibration. The same, we are now hearing from the great thinkers of the sciences, holds true in the laboratory. Hallelujah!

How vibration manifests in different frameworks . . . how we make the music . . . that is at the heart of our search for enlightenment and knowledge, as we resonate to a higher octave, in the key of life.

■ ■ ■ ■ ■

*From a very early age, I have known the gift of clairvoyant vision and it has led me through many a difficult moment, while creating others. That is the process of simply growing up human, dealing with everything that you believe comes at you—only to realize that it all comes **from** you . . . and that you create it all, at the soul level.*

We step forward, bold and curious; we step back, fearful and doubting. We turn to the left; we turn to the right. So is the dance of life. All of it is a reflection of music and that music, of itself, is nothing more, nothing less than the ensemble of all living things, in the great orchestra of conscious experience.

In the days of my peaking curiosity, I was ridiculed, like most metaphysicians, because I dared to express my ideas about extraterrestrial life, along with an array of paranormal pursuits and their resulting manifestations that people claimed were "ridiculous." But for the total support and encouragement from my true mentor, my mother, I found hostility at almost every corner. I dared to be a nonconformist and trusted the visions that swirled around in my head, from the earliest days of my memory. I spoke to spirits and they spoke back, and I got my validation from, shall we say, "otherworldly" sources.

I still do.

People with similar gifts have been locked away in institutions, dismissed as "insane." Others, in earlier dark times, have been burned at the stake as "heretics." I knew that, within this lifetime, things would change.

That time has come. Change—real change—is most undeniably upon us. The skeptics are now becoming believers; science is now embracing what it has for so long rejected; the poles of duality are being refined into cohesive, resonant beliefs—rich in many tones, many harmonies—of our individual existence, and the existence of a vast, unending, multiverse.

Always searching to share and explore new ideas with my peers, I decided to host a radio program back in 2007, which still airs today on BBS Radio, an Internet-based talk radio format. My show, Beyond the Matrix, *dedicates an hour every week to discussions with cutting-edge thinkers—whom I have sought out to interview—from every field of interest. I have spoken with some of our leading physicists, medical scientists, physicians, technicians, explorers, metaphysicians, and everything in between, intent upon bringing the new paradigms of our awakening civilization to a growing public.*

From these many interviews, some have touched me at a very deep and personal level. Whether reconfirming my understanding and perceptions, or challenging me to look past my own beliefs into a range of other possibilities, these visionaries have brought forward new and exciting perspectives on the human experience. They are dedicated to the positive outcome: believing in our capacity to right injustice, raise our vibration, heal the Earth, and most importantly, celebrate this hour of evolution.

They are some of the brilliant minds of our times.

Somehow, I felt those amazing dialogues should be brought out to a greater audience, to be shared with skeptics and believers alike, all united in one common goal: to raise our understanding of the meaning of our mortal and immortal existence, to seek a greater awareness of what is happening all around us, and to glean, from these, what our role in the greater Cosmos truly is. That is why this book was created and that is how it has come to be birthed, translated from the spoken to the written word.

From the illustrious master physicist Dr. Michio Kaku, co-creator of the String Field Theory of modern physics, to the warrior for the whales and dolphins of our seas, Hardy Jones, of BlueVoice, this collection of daring conversations challenges the reader to get out of the box of conventional thinking, to move beyond his or her own convictions, and to soar with the eagles, into the free-thought zone.

There, beyond the matrix, there are no thought police. There are no hate crimes. There are no rulers; there are no slaves. There are no victims. There is no fear; there is no separation. There is no right way, no wrong way . . . no definitive answer to the questions that plague our souls.

What can be found there, in the unlimited light of our incredible potential, is an infinite array of possible realities, in which we, vibrating strings on the instrument of life, play our melodies into the Music of the Spheres. There, we create the thoughts, which generate energy, and that energy forms the vibrations, which ripple across the waters of the One Mind. We listen to the grass grow; we celebrate the waves, as they break upon the shore. We hear the bird song, calling the flowers to bloom. We contribute our thoughts to the great collective of consciousness, moving forward, ever closer to the precipice where we leap, like the Fool of the tarot, into the unknown . . . knowing, trusting, that we have known it all along.

It is all written in the ethers, just over there, behind the veils of our three-dimensional perception.

And here we are again, leaving the cycle of ignorance to its own demise, standing tall, walking boldly forward—into the light of the New Dawn.

It is my hope and greatest desire that, from the insights brought forward from my exchange with these passionate human beings, whose visions for tomorrow are filled with hope and anticipation, your perceptions of life and humanity will be stirred to a higher level. Whether we are blinded, still, by the fog of our environment and emotions . . . or whether we feel we have traveled far along the highway to the stars, let us all recognize the journey up the spiral of light with celebration . . . of all that we leave behind us, and all that lies ahead.

That is our legacy.

Look . . . just up ahead. That is the Age of Enlightenment: turn left, at the next star, and you'll be there. . . .

The Physics
of the Impossible

■ ■ ■ ■ ■

Dr. Michio Kaku

Dr. Michio Kaku is a professor at City University of New York. Called by many the "Einstein" of our times, he is a brilliant physicist and a superb teacher for us all. He is the co-founder of the String Field Theory. He has written several books, and appeared in several documentaries, helping bring the complex theories of "new" physics into a language we can all understand. . . .

■ ■ ■ ■ ■

■ *Dr. Kaku, I have all of your books sitting in front of me . . . talk about a wealth of knowledge! Thank you for all you have brought into the light of human awareness.*

I'd like to begin this interview by telling you how I found out about you. Over a year ago, I was watching Discovery Channel, and I caught a program that, unfortunately, I came in on somewhere mid-stream, which showed a reenactment of a terribly distressed woman walking through a mall. People kept walking straight into her and she had no idea what was

*going on, other than a strong sense that she was somehow "invisible."
The show cut to you, in the middle of Times Square, where you were
explaining about other dimensions and how the people depicted in the
documentary were actually in another dimension.*

Would you like to elaborate on this world of merging dimensions?

Well, first of all, since I was a child growing up in San Francisco, I
used to stand for hours, watching the carp swimming in the Japanese Tea Garden. I used to imagine myself a fish, one of these fish
swimming in a shallow pond, and I contemplated how they live in
a two-dimensional world: they can only go forward, backward, left,
and right. Then, any one fish who dared talk about the world of
"up"—the world of the third dimension—the world of hyperspace,
was considered a "crackpot" or an idiot. I imagined a scientist there,
a scientist fish, who would say, "Bah, humbug! There is no world of
'up.' There is only what you can see; what you can see in the pond
is what there is. Period."

So I imagined reaching down and grabbing the scientist fish, lifting the scientist fish into the world of "up"—the world of the third
dimension—the world of hyperspace. What would he see? He would
see a fantastic world, where beings move without fins . . . a new law
of physics, beings breathing without water . . . a new law of biology.
Today, believe it or not, we physicists believe, and are trying to prove,
that we are the fish. We have spent all our lives in three dimensions
(moving forward, backward, left, right, up, down), and anyone who
dares to speak about unseen worlds, worlds beyond what we can
see and touch, is considered a crackpot.

■ *That would be me, by the way! I have had a lifetime of raised eyebrows
and outright laughter over some of my very deepest experiences in the
"beyond."*

Not anymore! The tide has turned rather dramatically. Now physicists believe that perhaps we can, in fact, access or begin to think about higher dimensions: dimensions that we cannot see; dimensions that are all around us, just like the world of "up" to the fish.

■ *What do you mean by the term "hyperspace"? In your books, you speak of ten dimensions, in a sense "layers" of reality. I am curious to know why you limit yourself to these ten dimensions and how you reach that conclusion.*

Einstein had a dream that he chased for thirty years of his life, and that was to create a "Theory of Everything," an equation one inch long that would allow him to summarize all the forces of the Universe and allow him to read the mind of God. However, whenever we try to combine the equations of gravity, the equations of light, the equations of ratio, we find that there is not enough room in three dimensions. There is not enough room to cram all these forces together. However, if you assume there is a higher dimension, in fact, up to ten dimensions, then all these forces simply, beautifully, elegantly, collapse into a single coherent theory, and that is what is exciting the world of physics. But Einstein didn't go far enough. Einstein stopped at the fourth dimension, but if you go to six, seven, eight, all the way to ten dimensions, then you get a very simple, elegant description of these higher dimensional universes.

We now think that the Universe is a soap bubble of some sort. We live on the skin of the soap bubble; we can't leave. We are stuck here like flies on flypaper, and our soap bubble is expanding—that fits all the data from our satellites. But we now believe that there are other bubbles out there, and what are these other bubbles floating in? What are these bubbles expanding into? If the Universe is expanding, then what is it expanding into?

We think that it is expanding into these higher dimensions, and for me this is very pleasing, because when I was growing up I went to Sunday school. I learned about Genesis and the beginning of the Universe, but my parents are Buddhists, who believe in Nirvana, which is neither a beginning nor an end. Now, we physicists believe we can meld Buddhism with Judeo-Christian Genesis.

We do believe that the Universe was born in an explosion, but that these Big Bangs happen **all the time.** They happen in a much larger ocean of Nirvana, and we even have a name for Nirvana now—that is: "Ten-Dimensional Hyperspace." We think that we have a beautiful picture of universes being born, growing, expanding, with other universes being born, parallel worlds and a larger arena that is eternal: Nirvana—the Nirvana of Buddhism; the Nirvana of Physics; Ten-Dimensional Hyperspace.

■ *What you are saying, then, is that we can go with the concept of the All That Is, That Ever Was, and That Always Will Be, and that the Big Bang will just become a manifestation of the perennial existence of the All?*

That's right! People have a hard time reconciling the Big Bang Theory with some theology, but this larger picture, this picture of the multiverse, gives us a very beautiful and elegant unification of many of the paradigms of the past, and believe it or not, this is the picture that fits the satellite data.

We have a satellite orbiting the Earth right now, called the *WMAP*. It has given us the age of the Universe, which is 13.7 billion years, and all the data fits this picture: this picture that we have—bubbles, like a bubble bath. Think of a bubble bath or a bubble that pops into existence and expands, baby bubbles, and bubbles collapsing into other bubbles. This is the new paradigm that is emerging from physics, and it is a paradigm that we have seen in religion throughout history.

■ *How does that correlate with religion?*

If you take a look at Genesis, chapter 1, verse 1, you see that it speaks of the instant when God created the Universe. That is consistent with the Big Bang Theory. In fact, there was even an archbishop of the Catholic Church who said that the Big Bang Theory is compatible with Genesis. But then we have certain forms of Hinduism and Buddhism that say there was no beginning and there is no end to this. It is timelessness; we have Nirvana. And so, we are seeing not only the melding of these two ideas—that Big Bangs happen within Nirvana—but that Big Bangs happen all the time.

Even as we speak, universes are being born and our satellite data is consistent with this picture—eternal Big Bangs, eternal inflation. In fact, one of my friends, Alan Guth, may eventually win the Nobel Prize in Physics for this picture that seems to fit all the satellite data. I think we have a very nice blend of science coming from satellite data, theory coming from Einstein's Unified Field Theory, and religious paradigms.

■ *One of the wonderful things about your work is that you manage to make the complexity of physics understandable to the layperson, for which we are truly grateful, because these are some pretty difficult concepts to grasp. That said, can you give us an overview of the String Theory (for which you are recognized as the co-author) and how that relates to quantum physics?*

The Pythagoreans of two thousand years ago were Greek philosophers who analyzed the lyre string, searching to find a mathematical ratio to music: the mathematics of harmony. Why do we have sharps and flats and thirds and fifths, and chords, and major and minor chords? Using the Greeks' understanding of mathematics available to them two thousand years ago, they were able to show that harmony obeys mathematical laws. They were so stunned by this realization that they thought maybe the Universe could be explained in the language of harmony. Well, they failed. They didn't know about atoms, they didn't have chemistry, they didn't have physics, and that idea never went anywhere.

However, now we have come full circle. Now we have too many particles. We have electrons, neutrinos; we have gamma rays; we have Higgs boson particles. When I got my PhD at the University of California, I had to memorize the names of hundreds, thousands of subatomic particles, with very exotic names.

Today we believe that all these particles can be summarized as nothing but musical notes—musical notes on a vibrating rubber band. If I had a microscope and I could peer into an electron, I would see that it is not a dot at all—that is the old picture—but, rather, it is nothing but a rubber band vibrating. If I can change the vibrational frequency, it turns into a neutrino; if I change it again, it turns into a chord. I change the frequency again, and it turns into all the hundreds of subatomic particles that I had to memorize to get my PhD in physics.

I would hope that in the future, when you get your PhD in physics, all you would have to do is say "String Theory" and you would get your PhD. Now we physicists believe that physics is nothing but the harmony of vibrating strings. Chemistry is the melodies that you can play on these vibrating strings. The Universe is a symphony of these vibrating strings and, therefore, what is the mind of God?

As I said, Einstein spent the last thirty years of his life chasing after a theory that would allow him to read the mind of God. Well, believe it or not, we now have a candidate—a candidate for the mind of God that has excited the world of physics. *Time* magazine, music magazines, and all the major networks have devoted stories to this— the mind of God. We physicists think it is cosmic music resonating through hyperspace: that is what we think the mind of God is.

■ *This is one of the reasons why I am so attracted to your work, because in my small contribution from the spiritual aspect, everything that I write about is about the music of the Universe as well, and when I read it coming from you, from a scientific perspective, I think, "Oh, boy, science and spirit are truly melding!"*

That's right, and next month, by the way, we are going to debut the biggest machine that science ever built, outside Geneva, Switzerland, called the Large Hadron Collider. It is seventeen miles in circumference and we hope to actually create the higher musical particles that we cannot see, because we are the lowest octave. Everything you see around us is little rubber bands vibrating in the lowest frequencies. We intend to generate the higher frequencies at the Large Hadron Collider.

Now the media, of course, has distorted the whole project. The media just say that maybe it will create black holes that can eat up the Earth, and that is just silly. That is such nonsense. We are not going to create any black holes in Geneva, Switzerland. What we hope to get is higher music. That is what this machine is all about.

■ *How does this relate to the question of the Music of the Spheres?*

The Music of the Spheres was an attempt by musicians centuries ago to be able to explain the motion of the planets, and the motion of objects we see around us, and to explain motion in terms of music. That idea never went anywhere in science, because along came Isaac Newton, who gave us the Laws of Motion, and that worked very well with our space satellites and with the planets. Then along came Einstein, who said that Newton didn't go far enough, that the motion of the planets can be explained, not just by gravity, but by space and time—the curvature of space and time. And now we realize, hey, Einstein didn't go far enough either; we have to go not just to space and time, but to hyperspace and through vibrating strings creating music—music resonating through hyperspace. That takes us all the way back to the Music of the Spheres. In some sense, we have gone full circle, in terms of our philosophy.

■ *Let me try and follow your thoughts through. Referring back to your metaphor of the rubber band and the concept of vibration altering the manifestations, aren't we of course saying that, if enough human beings on this planet raised the frequency of their thoughts, we could change the material essence of the planet?*

In this instance, I am not exactly sure how to change the frequency of these particles—one way to do it of course is with the Large Hadron Collider.

■ *Well, sure, this is the music, no?*

Yes, and another possibility is with the satellites that we have orbiting the Earth. By the way, I forgot to mention this: we now have baby pictures of the infant Universe ... baby pictures of the Big Bang. One thing that people have always accused us physicists of is that we have no graphic proof that there was this explosion that took place 13.7 billion years ago. But now we have photographs of it: photographs of the explosion itself. The photographs were taken in the microwave region. If you go to www.nasa.gov, which is the Web site of NASA, and you type in WMAP, you can actually see photographs of the explosion itself. So this is really gratifying proof of the basic picture. There was a Genesis; there was a beginning; there was a moment when the Universe suddenly exploded out in all its glory. We now have photographed it. The next question is: "What happened before this explosion?"

That is where we think String Theory comes in.

String Theory actually takes you before the explosion, so time did not begin at the explosion, which set off the Universe 13.7 billion years ago. String Theory predicts that there could be other universes—so, if the multiverse were like a bubble bath, then perhaps our soap bubble parted, or sprouted from another soap bubble, just like soap bubbles will cut in half when you are taking a bubble bath ... or perhaps two bubbles slammed into each other, to create a third bubble.

That's called the Big Splat Theory.

Some physicists, particularly my friends at Princeton, are pushing the Big Splat Theory; however, my friends at MIT are pushing the Budding Theory, that's called "Inflation." It suggests that our universe budded from another universe and we are taking bets right now! We don't know for sure which is correct. However, the point I am emphasizing is that the Universe and time did not begin at the Big Bang. We think that we can go before the Big Bang and we are actually going to measure this around 2014, when we send up a new generation of satellites that should give us a picture, not just of the explosion, but a picture of the instant of Creation, when our Universe was emerging from the womb.

Perhaps we will even find evidence of an umbilical cord—an umbilical cord connecting our baby universe perhaps to a parent universe. That, we haven't done yet. However, everything else we have a pretty good grasp on. We don't have proof of this, but we hope, by 2014, when we send up the next generation of satellites, to detect the umbilical cord that connected our infant universe to our current universe. All of this is pretty exciting stuff.

■ *It really is exciting stuff, especially because it is so in tune with some of the more far-reaching concepts that are coming out of the spiritual world. My books also talk about the astral cords of the sun. You are talking about the umbilical cord of the Universe. It is just wonderful.*

Another theme that has come out of religion and spirituality is the idea of a parallel world: other planes of existence. The Catholic Church, for example, has long believed that heaven and hell represent different astral planes, co-existing with our universe. We physicists used to snicker at this idea, but we don't snicker anymore, because all the evidence seems to point out that there could be other universes, other parallel universes.

Think of a radio, for example, in your living room. Your radio picks up one frequency, and you know, however, that there are hundreds of unseen frequencies in your living room—there is Radio Moscow, there is Radio Havana, there is the BBC, all these frequencies that you cannot hear. Now, why can't you hear them? You can't hear them because your radio is not in tune with these other frequencies (we say that is no longer "coherent"). That is the scientific term: it has "de-cohered" from BBC. You are listening to the rock radio station in America, you are not listening to BBC radio from London; you have de-cohered from BBC television.

We now believe that everything around us is vibrating at a certain frequency; we have shorter waves that are vibrating in our room; however, we also have other waves from other frequencies. Each frequency represents a different universe; so for example, perhaps in your very living room there are the wave functions of dinosaurs, the wave function of a world where Elvis Presley is still alive.

These are frequencies that we no longer interact with; we have de-cohered from them. This theory is called "Many Worlds," proposed by a physicist named Hugh Everett at Princeton in the 1950s.

People laughed at it when it was first proposed, but now we tend to believe that Everett was right, that there really are these other worlds. We have de-cohered from them; we are no longer in frequency with them . . . but we co-exist with them.

■ *Do you believe, Dr. Kaku, that some people are picking up these frequencies from other dimensions?*

I get stories from people who claim that. I don't know, because I would have to measure whether or not these people are attuned, but if you look at the physics literature, we physicists no longer snicker at this idea: the idea of Hugh Everett, that is "Many Worlds," is now one of the dominant themes in theoretical physics. You can go to any physics library on the Earth, look up "Many Worlds," and there you find hundreds of papers dealing with the quantum mechanics, the quantum theory of Many Worlds Theory.

Einstein didn't like it, but hey, on this question, Einstein was wrong. Einstein didn't believe in the quantum theory, but we now believe that the quantum theory does describe the world. That's why we have laser beams; that's why we have transistors. Without the quantum theory, there are no radios, there are no television sets, there are no laser beams, there are no satellites. Everything depends upon the quantum theory and the next extension of the quantum theory is Many Worlds.

■ *I read in one of the books where you talk about an "inter-dimensional lifeboat." Could you expand on that? I found it very intriguing.*

A lot of people come up to me and say: "Professor, all this is nice. You talk about higher dimensions, ten dimensions, but how do I visit these other worlds? How do I go to these places?"

Well, that is difficult.

First of all, religious and spiritual people have talked about meditating and going between universes. We physicists are more practical. We would like to have a machine that would take us, by pushing a button, rather than taking some kind of hallucinogenic drug or meditating. In my book *Parallel Worlds,* I even provide a blueprint of what such a machine would look like.

You have to understand that Einstein said, "Space and time are like a fabric," and that "time is like a river": a river that can speed up and a river that can slow down, as it meanders its way throughout the Universe.

We measure this with our satellites all the time—with the GPS system, for example, that you use in your rent-a-car. The GPS system depends upon time speeding up and slowing down in outer space. But the new wrinkle is that the river of a planet can have new whirlpools and that the river of a planet can perhaps fork into two rivers, in which case we cannot rule out the possibility of an interdimensional lifeboat. We cannot rule out the possibility of even a time machine that can take us between different rivers of time, different time streams.

Now this, of course, was always considered "crackpot," but Einstein himself realized that time travel was possible in his own equation. In 1949, the very first time travel solution was discovered in Einstein's equation. Since then, we physicists have discovered hundreds of different kinds of time travel solutions that allow us to go between universes. Stephen Hawking even calls them "baby universes"—little baby universes that allow us to go backward in time or between different universes.

To be fair, the machines necessary to go between dimensions would have to be huge. We are talking about a machine so big it has to be placed in outer space, but the very fact that we can even think about these things, I think, is quite amazing. Usually science fiction writers and fantasy writers talk about drifting through the river of time, or gateways: doorways to other universes. This is the stuff of the Twilight Zone, but we are taking this very seriously now, and we are even proposing blueprints for these kinds of machines.

■ *When do you hope that science will achieve such a monumental accomplishment?*

Well, it is not for us, in the sense that we are talking about huge amounts of energy... but one day, if somebody knocks on your door and claims to be your great-great-great-great-great-great-granddaughter, don't slam the door! It is conceivable that your great-great-great-great-great-great-granddaughter lives in a time when time machines are possible, and has decided to visit her illustrious ancestor, so don't necessarily slam the door if somebody claims to be a descendant of yours who doesn't exist yet!

■ *No doubt Steven Spielberg will love it, that's for sure.*

In fact, one of my friends, a physicist at Caltech, is even consulting for Steven Spielberg's next movie, which will be about time travel. Spielberg takes this very seriously. He actually consulted with a physicist to create his next movie, just the same way that Carl Sagan consulted with a physicist when he made the movie *Contact*. He actually used the latest physics to create the movie with Jodie Foster, which was about just that—contact with extraterrestrial civilizations.

■ *So when are you going to make a movie, Dr. Kaku?*

Not for a while. I am just shocked that I can even write a book that hits the *New York Times* Best Seller list! I am shocked by that.

■ *That's fantastic, and I am just going to name some of your most impor-*
tant books: Hyperspace—*which you have got to read;* Parallel Worlds—
another wonderful book—and we are going to get to the new book, Physics
of the Impossible, *in a moment. But before we do, I want to just jump*
back to what is going on in Switzerland. Can you tell us the date that this
is going to occur?

Next month, outside Geneva, Switzerland, a machine that is seventeen miles in circumference is going to be turned on. It is harmless,
it is based underground, so that there is no radiation that comes out,
and it is incapable of creating a black hole that will eat up the Earth.
I think the media got carried away on that one.

We turn on the magnets in May 2008. We get the beam going—
it may take a little while to get the beam up to speed—and what the
beam is going to do is to re-create a teeny weeny itsy bitsy little piece
of how the Universe was created, and again—Mother Nature creates energy far beyond this and nothing dangerous ever happens.

The Earth gets bathed, by the way, in cosmic rays that are much
more powerful than anything that we puny humans can re-create
on the planet, but we have created a machine that can give us a window: a window on some of the dynamics of Genesis. I am excited
by it, because it may help to prove the String Theory.

Some people say that String Theory cannot be proven, because
you have to create a baby universe in the laboratory. Well, we are
not gods; we cannot create a baby universe in a laboratory. But the
next best thing you can do is to create higher levels of music. We
hope to create what are called "sparticles." Sparticles are super particles; they are projected by this little rubber band that vibrates in
five dimensions, and we are the lowest octave of this rubber band.
But there are higher notes—we haven't seen them yet—and that is
why we want the machine at CERN, outside Geneva, Switzerland,
to create some of these higher vibrations.

If that occurs, it will send shockwaves throughout the world of physics. It will help to quell the cynics who say that the String Theory cannot be tested, because we cannot create baby universes in the laboratory. We can do something almost as good, which is to create sparticles, which we think are the next set of vibrations of a vibrating string.

■ *This is exciting and also very interesting that such a monumental endeavor is taking place, just years before the famous 2012 transition. Many people are really manifesting a lot of fear about 2012. I would love to ask you if you have any feelings or perception about what we are headed toward.*

Well, let me say a few things about that. I have looked at some of the Mayan translations, and if you go into the calendar of the Mayans, they talk about cycles, and these cycles, of course, take place over thousands of years. We are headed for the end of one cycle. There are two ways you can look at it. One is, of course, that perhaps we are at the end of a cycle, the end of the Earth. But the way the Mayans themselves looked at it, I think, was as a beginning. It is a cycle; it is a rebirth, so it does not necessarily mean the end of everything; it simply means the rebirth, a recycling. We have gone through previous cycles in the past and nothing happened, and we are going to go through more cycles into the future. I look at it more as renewal rather than the End.

■ *So do I. I look at it as an incredible archetype of our individual process of reincarnation.*

Yes, so if we look at it, rather than from the dark eyes of cynicism and depression, I think we should look at the Universe as glorious. It is absolutely amazing that the Universe exists at all! It is a splendid, fantastic place that we live in, and it is very hard to believe that a cycle can end just all of a sudden, where most cycles represent rebirth. That is why, if anything, I am quite optimistic about the way the world is unfolding. The world, of course, is facing many challenges at the present time, but the human spirit, I think, is indomitable, and I think we will overcome these hardships. We have done it in the past, and I think we will do it in the future.

■ *I am so glad to hear you affirm that. We need voices of positive thought and celebration, despite all the darkness and difficulties on this planet right now. I agree, wholeheartedly, that the human spirit cannot be suppressed and that it is when we face our greatest challenges that we have our greatest opportunities to rise above them.*

One good thing about looking at the world through cycles is that you tend to look at the previous one and you begin to think, "Where were we? What obstacles have we overcome in the past?" and that is a guide for the future. Instead of seeing gloom and doom, I see areas where we can begin to reevaluate where we have been. Unless you understand where we have been, you cannot understand where we are going to go, and the very fact that our looking at our history through cycles forces us to reevaluate the past, I think, gives us hope, in terms of being able to forge ahead into the future.

■ *What do you think is going to be around the corner for us? How do you foresee this transitional period ahead of us, with all the threats of so many dark things that we don't need to list them, contrasted by the hope for so many positive evolutionary changes? How do you see the next few years playing out on this planet?*

Let me talk to you about how the next century might play out, because I do believe in birth and rebirth. Physicists seriously look for evidence of intelligent life in the Universe, very seriously, and we also have to very seriously analyze the question: "What will these civilizations look like once we have made contact with them?" For us this is not an academic question: we have telescopes, we have satellites, and one day we **will** make contact.

Some physicists have theorized that civilizations occur in three types: Type One, Type Two, and Type Three.

A Type One civilization would be truly planetary: they have a planetary culture; they would have mastered the weather, hurricanes, earthquakes; they would have cities on their oceans. A planetary civilization fitting this description is about maybe one hundred years more advanced than us.

Then we have Type Two: a Type Two civilization has actually gone into the stars now and they actually control stars . . . a few neighboring stars. They colonize a few neighboring solar systems; they control the energy output of an entire star.

Then we have Type Three, which is galactic—they actually control the energy of not just one star, not just a hundred . . . but billions of star systems. That is Type Three, and if any civilization has ever visited us they would probably be Type Two or Type Three.

Now, if you take a look at us, what are we on this scale? We are Type Zero.

■ *Okay, so how do you define Type Zero, which is sounding like a rather un-evolved civilization?*

Okay, let's talk about Type Zero. Type Zero gives us energy from dead plants. They still have the savagery of the jungle, the savagery of sectarianism, ignorance, poverty, disease. They can only dream of being a planetary civilization, but if you get a calculator and just do the calculations, we are about a hundred years away from being a planetary civilization, and I see the birth pangs of a Type One civilization every time I read the newspaper.

For example, what is the Internet? The Internet is the beginning of a Type One telephone system. For the first time in history, we can actually see the outline of what a planetary telecommunication system would look like. The Internet is the baby, the germ, the seed, of a Type One telephone system. What language will they speak? They would probably speak English. English is already the language of the elite. Everywhere I go to speak at conferences, the elite businessmen, politicians, and scientists all speak English. That will be for the middle class, a hundred years from now. The middle class will all speak English, as well as their own language.

What is the European Union? The European Union is the beginning of a Type One economy—and why do we even have the European Union? To compete with NAFTA, that is, the United States. So we are seeing the beginning of a Type One economy, and everywhere I go (and I listen to music and watch TV), I see the beginning of a Type One culture. Young people love rock and roll; they love rap; they dance to the latest beat. The elite have high fashion, which I think is pretty much uniform now throughout the world. We all love movies—that is also giving us a Type One culture.

We see the beginning of a Type One culture everywhere we go, but there is also a backlash, and this is what I am afraid of. There are forces that don't want a Type One planetary civilization, because it is multicultural; it is enlightened; it is progressive; it believes in equality; it believes in progress in the future. These are terrorists. Terrorists believe in Type Minus One civilization—a civilization dominated by dogma, by rigid religious beliefs, and by a very rigid hierarchy, where males are at the top, and women are at the bottom. That is the terrorists, who don't believe in a progressive planetary Type One civilization.

It is not clear that we are going to make the transition from Type Zero to Type One. Some people think that, maybe in 2012, we will make the transition to Type One, but I tend to think of 2012 as nothing but a renewal, a new beginning. Where are we headed for? Where is the Age of Aquarius that people have talked about? For me, that is a Type One civilization—planetary, tolerant, progressive, scientific, democratic—I think that is where we are headed, unless we blow ourselves up first!

■ *I would like to add just one thought to your elucidation of these differentiations of planetary levels and add that it is not necessarily only the so-called "terrorists" who don't want the evolution of our species. There are other forces at work besides the so-called terrorists, wouldn't you agree?*

Oh, yes, I would agree.

■ *I got that in there so that we don't typecast anyone.*

Right, and of course, we have nuclear weapons, we have proliferation problems, chemical weapons, biological weapons, the greenhouse effect, global warming. We have many forces which may prevent us from becoming a planetary civilization, so I see that there is a race against time. That is the paradigm that I see.

On one hand, we have the forces of enlightenment, education, the forces that are creating a prominent multicultural, planetary civilization. On the other hand, we have the forces of illiteracy, sectarianism, fundamentalism, racism. We have a dark side of us that came out of the forest. We still have that old, old brain of ours that had to slug it out with everybody else in the swamp and in the forest, and we still have that with us. It is not clear which one will win out, but I tend to think that, overall, we will survive and we will make it past 2012, and eventually make it to a Type One civilization. I am optimistic.

■ *As am I! That said, tell us a little bit about what is the overall significance and purpose of your book* Physics of the Impossible.

Arthur C. Clarke, who just passed away a few weeks ago, made a very famous statement and that was, I quote: "Any sufficiently advanced technology is indistinguishable from magic."

Now, think about magic.

Magic is being able to disappear, to become invisible, to reappear someplace else. Magic is the ability to turn objects into other objects, and make them disappear and then reappear someplace else. We are going to have much of these abilities in the coming years—like invisibility, for example. Two years ago, we made a huge breakthrough in invisibility research at Duke University and also at Imperial College in London. We were able to take an object and actually make it disappear, at least when you shine microwave radiation at it. At Caltech, just a few months ago, they showed that visible light can also act in the same way. And I think that in the coming decades, we will be able to take an object and make it invisible—so, Harry Potter, watch out!

■ *Isn't this the nature of the famous "Philadelphia Experiment"?*

Well, in some sense, yes. We are now able to show that light can wrap around an object, and then re-form at the other end, which we once thought was impossible. I teach optics to my students here at the university, and I always teach them that invisibility is possible because light, unlike water, can wrap around a boulder and re-form at the other end—like a river.

■ *How do you do this?*

There is a new substance called "metamaterial." It has lots of impurities in it, which we didn't think about before, but these impurities kick the light beam in such a way that it can wrap around an object. Take Harry Potter and put him in a cylinder. The light that hits the cylinder wraps around the cylinder and then re-forms at the other end. Proof of this principle was demonstrated two years ago. This shocked the world of physics. Every single physics textbook is wrong on this question.

■ *. . . But this is amazing! This is the "physics of the impossible," right?*

Yes, this is the physics of the impossible! Now, another "impossible" thing is to disappear, like a magician, and reappear someplace else. We physicists call this "teleportation" and at the atomic level, we have been able to do that in the laboratory. We have been able to make a particle of light disappear and have it reappear a hundred miles away, from one Canary Island to another Canary Island. Next we hope to do it to the Space Shuttle. We want to take a particle of light on the Earth, make it disappear, and have it reappear on the Space Shuttle. After 2020, we have plans to do it to the moon. There is no limit to this. We can actually make an object reappear on the moon. Now, for atoms, not just light, we have actually done it with cesium atoms and beryllium atoms—this has been done in the laboratory, and I suspect that, within a decade or so, we should be able to make a molecule—maybe DNA—disappear and reappear someplace else, and maybe a virus—maybe a cell—in the coming decades.

To do this with a human being is not yet possible. Let's be frank about this: a human consists of fifty trillion cells. We cannot teleport a whole cell yet, but we are getting there, and I think we have to think about the philosophical implications of this. When you are teleported, you actually die; you disappear. You dissolve in the process and then you reappear someplace else. This is depicted in movies, like *Jumper* and *Star Trek*. You have just seen Captain Kirk die in one chamber and yet here is this impostor, Captain Kirk, over there, saying: "I am the real Captain Kirk; I have his memory, his genes, his neural circuits, his personality quirks."

It makes you wonder about the **soul**. What happened to his soul, if you just saw him dissolve and then reappear someplace else, with all his memories intact?

■ *. . . And that is the question that I and other metaphysicians are concerned with.*

We physicists are scratching our heads about this. I mean, what does it mean to be you, yet your atoms all dissolve, so that you are no longer here, whoever you were . . . you are gone? Who is this other person over here, who is just like you, has exactly the same thought patterns as you, the same memory, and claims that he **is** you? This is an academic question that, in the coming decades, we are going to have to deal with. Is there a soul? We are going to face this question in the laboratory.

In the book *Physics of the Impossible,* I show that many of the technologies that you see in the movies (like *Harry Potter, The Terminator, E.T., Star Wars*—many of the Hollywood blockbusters) are, in fact, based on technologies that we physicists think might be achievable in the future. I can't give you a time frame—decades, centuries, years, millennia—for when we can begin to attain some of these technologies.

■ *Doesn't it seem to be sooner than that? Let's look to* Star Trek *again, with those handheld communication devices. A colleague of mine says Nokia modeled their product around the* Star Trek *props.*

Yep, we have those already, right!

■ *We have those already and they look and function exactly the same.*

Right, and look at telepathy. You know telepathy used to be considered the backwash of fortune-tellers and Las Vegas magicians, but we physicists have actually been able to duplicate limited forms of telepathy today. For example, at Brown University, they have taken stroke victims who are totally paralyzed, put a chip in their brain about a quarter the size of a penny, and connected the chip to a laptop computer. Within three hours, we trained these people to read e-mail, to type, do crossword puzzles and play video games and surf the Web.

■ *That is really spooky, Dr. Kaku.*

... but you see, think of all the people who are vegetables, trapped inside a paralyzed body, who cannot do anything ... and now they can e-mail.

■ *Yes, but let me interject, think about all the healthy people who could conceivably become trapped inside a computer mind!*

Well, what happens is that people in the future will have a choice. They **will** have a choice. Some people may actually want to have a microscopic implant and they will be able to surf the Web just by thinking about it. This is something that is well within the realms of possibility.

■ *Will they have a choice? Let's hope that they think a little bit beyond that and think about the significance of having a chip inside them and what that would imply.*

Now, if you want to know something spooky, here is something spooky. We can use MRI machines, brain scans, to look at the brain patterns of somebody telling a lie. When you tell a lie, it takes more energy than to tell the truth. To tell a lie, you have to know the truth, invent a cover-up, and then look at the consequences and consistency of the cover-up. This is a lot of energy and it is very easy to pick up on an MRI scan. In the future, we may actually be able to see the outlines of thought and emotions, just by reading the MRI results of a brain scan. This is going to go to court this year, by the way.

There is an insurance company that denied the insurance claim of a person whose house burned down. The insurance company said, "You burned that house down yourself; therefore, we are not going to pay you a dime."

This guy got very angry and replied, "I am going to go to court and have my brain scanned in court and I will prove I did not burn down my house."

In the future, this may definitely go to court. We may be able to solve crimes or whatever by reading the minds of potential criminals.

■ *Do you think this will take us to being a Type One planet?*

Well, the ethics questions are definitely raised, of course, because once you are able to zap people across the room, what happens to the soul? If you can read people's minds, what about privacy? Do we have privacy? Do we want people poking around in our thoughts, like in a science fiction novel? These are questions that we are going to have to confront. Certain things are coming whether we like it or not, and this book, *Physics of the Impossible,* tells you what is coming down the pike ... whether you like it or not.

I also talk about the possibility of starships in the coming decades. NASA is already beginning to think about what it would take to send a ship to the stars. In fact, I am a consultant for NASA. I have to review some of these proposals for NASA to build a starship, which, of course, is decades away, not anytime soon, but we are looking into the possibility of making contact with alien civilizations in outer space.

This is no longer just something for the movies; we scientists are looking at this very carefully. Paul Allen, the Microsoft billionaire, has donated twenty-six million dollars to building a new array of telescopes—radio telescopes—to eavesdrop on conversations between alien civilizations, so this is big business now. Billionaires are beginning to put money into this field, and some people are even making claims now that, within twenty-five years, we may make the first real contact with another intelligent civilization.

■ *I would like to suggest that it is going to be way before then and maybe sometime in the future you will remember this conversation, but I think it is going to be a lot sooner than twenty-five years.*

It could be ... it could very well be sooner.

■ *I do want to ask you, what "impossible" invention would you love to see created? What do you think humanity needs the most?*

Ah! What do we need the most right now? I think we need a way in which we can solve all these ancient conflicts that have been with us since the start of humanity, and find some equitable way forward. Most people, of course, would rather slice the pie thinner and thinner: more money for health care; less money for this ... the juggling that takes place in the political arena. I would hope that we make the pie **bigger.** Instead of fighting for smaller and smaller pieces of the pie and getting into civil wars and invasions, I would hope that science and technology can give us a bigger pie, so that we don't constantly engage in all these sectarian fights over cutting up the pie.

But if I were now to think about a machine that would really liberate us, perhaps it would be to build a time machine to look into the future. We might be able to see how things play out and learn lessons from this, and perhaps even one day change our own future.

■ *Wouldn't that take the joy out of the discovery of it all?*

Yes, but it would also take the agony and the suffering out of it as well. It's a trade-off.

■ *All right, here is what I would like to suggest: as you, in your incredible life and mind, explore ways to create the impossible, perhaps one of these machines that will come to us starts with your invention in Switzerland, where you create a higher field. I do also believe that everything is music, and hopefully we will see a higher octave on this planet and throughout outer space and the Universe.*

Yes, I would hope so; I would hope that I would fulfill Einstein's dream of being able to read the mind of God. I think we are all now approaching some really cosmic questions that we can begin to solve in our own backyard, right? It is thrilling that we are going to be alive to see a lot of these questions being answered with our breakthroughs in technology.

Hopefully, String Theory will guide us to this path.

■ ■ ■ ■ ■

Dr. Michio Kaku's work can be followed through his books and Web site: www.MKaku.org. His new book, *Physics of the Impossible: A Scientific Exploration into the World of Phasers, Force Fields, Teleportation, and Time Travel,* has already hit the *New York Times* Best Seller list.

Other books include:
    Parallel Worlds, 2006
    Einstein's Cosmos, 2005
    Visions, 1998
    Beyond Einstein, 1995
    Hyperspace, 1995

Where Science
and Spirit Merge

■ ■ ■ ■ ■

MaAnna Stephenson

MaAnna Stephenson is known as a visionary thinker, one who fuses science and spirit. She began her writing career with a short story triggering "soul memories" and began her self-education in the fields of technical, scientific, and New Age thought, exploring ancient mysticism and the rational sciences with equal emphasis. After a five-year preparation period, she was initiated as a shamanka. Her training for this initiation helped her to reconcile the rational sciences and the intuitive arts. This process has culminated in the writing of her book *The Sage Age: Blending Science with Intuitive Wisdom.*

■ ■ ■ ■ ■

■ *MaAnna, your book is an encyclopedia of spirit and science. How long did it take you to research it? A lifetime, I would imagine.*

I wrote it for myself as a reference tool. Then, my lifetime experiences definitely came into it, but as far as actually sitting down to research the topics for the book, it took about four and a half years.

■ *That sounds like the "scientist" talking, but for sure the intuitive voice was just flowing in on the wings of cosmic intervention.*

Oh, yes, every day, through the shamanka work! I think most people don't understand what "shamanka" is. It is the feminine term for "shaman." Most people think of shamans as medicine folk, but not all of them are. The basic meaning of "shaman" is someone who journeys to other realms and brings back information to anchor here. That is basically what I used that for, during all of the research and the writing of the book, because there are just mountains of information to sort through. It is important to be able to pick out what is most meaningful and help connect the dots between all of the topics, because there are so many tangents out on all this "specialties-type" information that is in the book. I try to pull together all the things that can help people connect, to help them see that science and spirit are actually investigating and talking about the same sorts of things—maybe just using different language that the other side doesn't understand or relate to.

■ *That is what is so exciting about this time of our evolution. On my show, I have had quite a number of people that are right on the same wavelength, like Michio Kaku—I am sure you know him. I remember when I first saw Michio Kaku's video, which he opened with the statement: "On the one hand I am a Buddhist; on the other hand, as a physicist I believe in the Big Bang Theory, so I have a conflict!"*

I was delighted to hear this from such a noted physicist. I try to bring the dual nature of all information forward, because I am myself a channel, and I too am working with information from other realms. I am also a shaman, so I am always looking to help people—even the skeptics—to find their way over to the spirit side, while helping spiritual people get the grounding that we need to be credible, and more—to be connected to Mother Earth.

Yes, because a lot of this work requires all of it—the science and the intuitive arts both require much discernment.

■ *Absolutely. I think now more than ever, as the veils are lifting and the dimensions are weaving through each other, that we are going to have to be very grounded at this time of shifting, and this is something that I am really stressing quite a lot.*

I found your biography very significant. You talked about your father being an engineer and its impact upon you. Could you share that with us a bit?

Yes, he very much had an engineering mind. He could sort out just about anything. No matter what you put in front of him, he could figure it out, and what he taught me was to go through things in a methodical fashion. He taught me how to break big things down into little pieces and to understand the little pieces, and then understand the whole of it: how everything worked in harmony together.

That was really a wonderful learning process.

He was also an artist and a musician, so I got to see his creative side and his expressiveness through that. He was able to balance one with the other. That was really a wonderful upbringing to be around. My mother is still singularly the most creative person I know, because she is not bound by the laws of physics. She surely doesn't have to concern herself with how things "work." She just knows what she wants and she can **see** it. She is very visionary and creative in that way . . . while it was up to my father to figure out how to make it work.

■ *You were fortunate to have the "best of both worlds" in your parents!*

Indeed I was! I lived right in the middle of that and helped with both sides: I got to see someone create things, and I watched how to make things manifest into the world. It was really an incredible upbringing. All the women in my family, all throughout my extended family as well, are intuitive, so growing up I simply thought that that was what it meant to be a woman: to be female. I was probably close to entering into junior high before I realized that not all other females were necessarily intuitive, in fact. It was a very strange thing to awaken to and then figure out, "Oh, you mean everybody is not like us?"

I know what you mean. When I was a young woman, if I would watch something painful enacted on the television, or if I would see anybody get hurt, I would tell my mother how I could feel the "psychic pain" up my back, down my leg, in my head, and throughout my body.

One day, she asked, "Honey, what exactly do you mean by 'psychic pain'?"

It was such a shock to me that she didn't know, and when I finally talked about it to other people, none of them knew either! You grow up with varying levels of sensitivity and so-called "supernatural gifts," but sometimes you have no clue (until someone questions you) that they aren't common to everyone. That realization and the surprise of its discovery were very much a determining factor for what set me off on the path I chose early on, to serve as a healer, which I did for thirty years. I can feel the pain of others physically in my body, mirroring their own, which, of course, is an amazing tool when you are healing others.

So, I do understand what you mean and I am sure this was something that really triggered you, when you came to terms with the fact that not everybody is an intuitive . . . and that you had this gift of vision.

Yes, it most certainly did. It has shown me a lot of ways of moving in the world, and it has really helped me to balance one thing with another, because of all the different things that I am into. If it is creative, I am there. With music, with engineering, with everything I do,

I have found that one thing helps balance another out, helping to give me perspective and a diversity of knowledge.

I believe that diversity is really important. We have overspecialized to the point that so very often, the right hand doesn't know the left hand is there . . . much less what it is doing! People don't realize how one thing affects another, because they don't look beyond whatever it is that they are concentrating on, to see a bigger picture.

I think now is the time to "be" Leonardo da Vinci. He was willing to look at everything from every angle, and that is how he was able to bring such wondrous things into manifestation: things that had never been thought before him. He was willing to be an artist, just as he was willing to be an engineer, and he was willing to combine them, finding great value in what he could learn through the rational, and what he could learn through the intuitive.

He had the capacity to bring all of those experiences to bear into manifesting something that had never existed on the face of the Earth. With all the specializations that we have now, it is time for that kind of appreciation of knowledge and experience, drawing from all aspects of learning, to help give a check and balance to what we are doing, so that we can find a different way to move forward.

■ *I like that idea of "being Leonardo da Vinci," a true visionary and seeker of truth, probably in many lifetimes. He is a powerful icon, in which we can find much inspiration of what we can achieve.*

As for the material: in the book, you present a fascinating concept about the "body antenna"—if you could explain that for us?

That is actually what motivated me to do the research in the first place. We are learning how the body is actually a multifaceted antenna system that receives lots of different kinds of information. That includes your physical body, all the way from your bones down to the cellular level of your existence. The brain processes the physical information that is coming in, and then there is also the subtle energy in your body that interprets information coming in from that way, as well. The mind-body antenna coordinates the whole thing and takes in information from everywhere.

Much of my book is devoted to that concept, showing how physical antennae work like radio broadcast and TV antennae. It also delves into perspectives on how things like yoga and tai chi, which move our bodies through different ritual positions, help facilitate receiving these frequencies. We position ourselves just like we change the rabbit ears on an old TV set, and move through these different energy waves, like a radio scanner that moves through different bands of frequencies to pull in information.

■ *It will be interesting also to talk about how interacting with other people, with different body language, can create the harmonics to receive their vibration.*

When you mirror someone else, you are setting your body antenna exactly as theirs and, in broadcast terms, that is how antennae receive. When you shape your antenna exactly like the waveform that is coming in, that is the best reception that you can have. When you put your body in the same length and same orientation to the earth plane as the person that you are dialoguing with, it brings you into more synchronistic thought with that person.

Good salespeople and counselors know this, and they mirror the other person for that reason.

■ *Other chapters talk about waves and sonic electromagnetic waves—all subjects that I find really cutting edge to our experience right now. Can you expand on these aspects in human experience?*

Yes, there is some confusion around this basic aspect of physics. People talk about "frequency" and "wavelength," but they often don't really understand what frequency and wavelength are. These words can have radically different meanings, depending on whether you are talking in scientific terms, or whether you are talking in spiritual terms. There is also confusion in the concept of the "continuum of sound frequency" and the "continuum of light," since some people believe they are part of the same spectrum and part of the same continuum. But they are not; they are entirely different sorts of light.

Sound is a physical entity, whereas light isn't. It is made of particles of no mass: it is not even a physical entity. It is just pure energy. If you come way out into an extreme broad scope of things, of course, everything is **energy** . . . just pure energy.

■ *In fact, I was going to say that the metaphysicians are going to say "Hold it!" here.*

Right, you have to come out to a very, very broad place to be able to put those in the same continuum, but science deals with them as completely different aspects that actually have nothing necessarily to do with each other. They just both happen to be measured in frequency, in wavelengths; that is all they have in common, in scientific terms.

It is like measuring anything else: just because we measure two different things in "feet" doesn't mean they are the same thing. If you measure a tree and then you measure a stone, they are not the same thing just because you measure both of them in feet or meters.

■ *What are you saying then, with regard to the basic difference between sound waves and light waves?*

A sound wave is a physical shape that is being moved through air. It is physical molecules, being shaped by the movement of a sound wave. Light waves are nonphysical; they are made of particles of no mass. They have absolutely no mass, and it's an electric wave and a magnetic wave that are propelling each other. Sound waves only go so far, and then they run out of energy and become absorbed. Light waves have no such limitations. They just keep going, until they are literally broken up by something. They're self-propelling waves. That's why we can see light from distant stars. We certainly can't hear sounds from them. Sound can't travel in the vacuum of space, either, because there is no air to move. So it's a physical entity, whereas light travels....

■ *So does sound require air to be able to move through?*

Right. All these science fiction films that you see, where you see the out-side of spaceships with their roaring engines, are just that: fiction. If you were actually outside the spaceship, you would hear no such thing. There's no sound in space: it's dead quiet. There's no sound of any kind, but there is light. So that's one of the main differences between sound and light. As I said, sound requires a physical medium to move through, and that can be a solid medium, like the Earth. When an earthquake rumbles, you literally have sound waves that go through the planet; part of that rumble is the physical wave moving through.

■ *Can you explain to us why that is? I mean it's a bit simplistic for you, probably, but not for us nonscientists. It would be interesting if you could explain why sound requires a physical medium to move through.*

It is a physical wave and it requires a physical medium. It moves through matter; it is transported across matter. The energy is trans-ported from one molecule of matter to another molecule of matter, and these molecules bump into each other and displace each other. And that's a physical wave.

■ *So let's say a horn in the spaceship is beeped and it gets sent out into the vacuum of space. What happens to that sound wave?*

Nothing. There is no sound wave.

■ *Does it get absorbed?*

No. There is no wave. There's nothing to propagate. There's no medium to move through. There is no medium of any kind in space to move through. That's one of the controversies about light that is still ongoing and has been since antiquity. We think, like in the case of sound waves, that light needs a medium to propagate through.

It was thought for a very long time and by some people still that space is not a vacuum: that it's not empty and that there is an "ether." They believe that since light moves through it, it has to have a medium to move through, and that was an ongoing investigation even in Einstein's time. And it's not dead yet. There are still people looking for the ether, even though nobody has been able to find it. Nobody has been able to prove it. They're still looking for it, to prove that light needs something to travel through.

What Einstein gave us, with the Special Theory of Relativity, was the confirmation that ether is not necessary to propagate from distant places through space. So science settled there, believing that if it wasn't necessary and nobody could find it, then, for the most part, there was no reason to look for it. That's pretty much where it lies today, although there are researchers who still believe it is necessary for light to propagate and so they are searching for it. If they can prove the existence of ether, it will change everything we know about cosmology. So that's why some people are looking for it.

■ *... Let's say we call that the "economy of research."*

Yes, absolutely. There are people spending billions upon billions of dollars trying to find something that's totally theoretical and that some quantum equations absolutely preclude ever finding. But that doesn't keep them from spending the money to insist.

■ *Well, we can probably give them some suggestions on better ways to spend it ... but that's too political for our intentions here. Let's talk about your fascinating presentation of how the body is a crystal lattice. I'm sure a lot of people will be fascinated to hear your take on that. Although a lot of metaphysicians have been talking about the crystalline nature of our makeup for eons, I like what you say.*

It is really the key to new imaging technology for the body. One of the things that I hope to be an active participant in helping to bring about in the world, is the little device called the "tri-quarter," which was in a *Star Trek* episode. Without touching anything, it could measure anything it was directed at (whether it was just energy and light) and could determine how those things interacted with matter. It could measure that and tell you the composition of it, and what was behind it, just as it could look inside it. They had a medical tri-quarter that would do the same thing, and report the condition of the body at an operational level.

I really want to see this developed in our lifetime, and we're really a whole lot closer to it than most realize; we're just one novel idea away. The technology actually already exists. Einstein gave us "$E = mc^2$," which is applied to all existing technology—the springboard for all the latest technology, like lasers. To be able to image the body through something like an MRI, which is a magnetic resonance scanner, requires using the crystals in the body. It uses the electrons in the body. With the magnet, it induces everything with a magnetic field:

it turns everything in one direction, then it turns the magnet off and takes a picture to see what turns back the other way. It uses the crystals from the body as an antenna to broadcast what's going on inside the body. That's how the thing works.

Most of the other imaging technology available, like the CAT scan, works on the same principle. Your body, bones, tendons, cartilage, intestines, even your teeth, are full of piezo-crystals, which convert electrical energy into mechanical energy and vice versa. They're the same things that are used in guitar pickup, electric guitar pickup, and phonograph needles. When you stamp your foot, you generate mechanical energy. You're stressing your bones. You're pushing them together with that impact and that's transferred, through these piezo-crystals, into electrical signals that run up your nervous system to your brain and say, "Hey! Can you come back here and reinforce this bone, because somebody is stamping on it?"

That's how we build our bones through weight-bearing exercises, walking, and other physical activities. The piezo-crystals facilitate that sort of thing. Also in your gut area, where your intestines are full of piezo-crystals, they convert electrical signals from your brain, from your nervous system, telling the body it is time to eliminate and detoxify. That changes into the mechanical movement that helps you process things out of your body. This is part of what makes the body a crystal latticework.

We also have liquid crystals all through us, just like on your computer screen, the flat screen, the liquid crystal display, the LCD: liquid crystals. You're actually one big crystal, which converts electrical signals into mechanical energy.

Research has been ongoing since the 1960s to heal bones with magnets. There are a number of magnetic healing techniques being used now: people can't quite figure out how they work; they just know they do, because the body responds to them. Part of the healing process is accomplished by the magnets' work with this crystalline latticework of the body. When we tune our body with the body antennae, when we place it in a different orientation by maybe sitting in a lotus position or doing some of the other yoga postures or tai chi postures, we're literally orienting this crystal latticework— just like a radio receiver uses crystals, when oriented in certain ways, to pick up and change frequencies.

■ *That's just amazing. I mean, as metaphysicians, we do teach people that the crystal is within you and that you do not need a crystal outside of you for energy alignment. You are the crystal. But to hear it explained scientifically is very exciting.*

Certainly. Anyone who works with crystals is only just enhancing that frequency. And it is not limited to quartz and other minerals: you also find it in other materials, like wood and silk. They're full of these crystals as well. So, folks who work with these tools in their healing practices are only enhancing the delivery, the transmission, and the reception of a frequency; possibly the crystal helps them focus their intention as well.

■ *Do we have these piezo-crystals at every juncture of the body?*

Not just that; they're in our bones. I mean, that's part of what makes up our bones. It's what makes up our intestines, our teeth, tendons, cartilage; they are a part of our actual structure.

■ *How does this relate to the work of Dr. Emoto and his water crystals?*

Well, you know, what he's working with is slightly different. He's freezing the water and then observing the layout of the crystal, to see if it forms in a cohesive pattern. The piezo-crystals that are in your body have a cohesive pattern at all times. They're always cohesive.

■ *Well, the nature of crystal would be that it does have a distinguishable pattern, right?*

Right, but the crystals that he's working with do not; they're a different constitution. They're not made up of the same things. What he's looking for is a guiding principle behind the formation of that crystal from water. He's looking at something a little different than a quartz crystal, which already has a guiding principle and has a predetermined growth pattern.

■ *Sure. But what my question for you is this. Since I do really embrace the idea that the water crystals are being imprinted with prayer or meditation and positive thoughts, they seem to transform themselves into patterns that reflect a higher vibration. Wouldn't this be a case for helping people understand how belief, mind, prayer can affect directly the body?*

Absolutely. And there is still a big question mark about what thoughts are and how thoughts manifest into reality. What exactly is a thought? Science is saying that they're developing answers to this question. This is not in the future. This is right now.

Scientists have developed ways for us to map the brain in such a way that we can identify the very specific places in the brain that have electrical activity when we're doing something, like moving an arm. They've been able to map these regions of the brain, observing how we can literally move an object by thinking about moving an object. Because we're thinking about moving our arm, the electrical activity in the brain is activated. So we don't physically have to touch anything to move it, if it's wired to receive a signal. What science is saying, and what's being popularized in the media, is that we're moving things with thoughts.

This is not correct. We're moving things with brain activity. Nobody is measuring our thoughts, because nobody can measure a thought. What they can measure is the shadow, the mechanical shadow of thought activity: the electrical firings in the brain. When a thought goes out, nobody can know where it goes; nobody knows how much power it has; nobody knows what that thought was. If you send an intention to another person for their safety, I don't think it can be measured. You can measure the effect of it on their safety. You can measure the brain activity—but you can't measure the thought.

We're not there yet, but that's what science is mistakenly saying. They're using those words to suggest that we're measuring thoughts. And that's not the case at all. So, there's a gap between how intuitive practitioners work with intent and understanding how it works in the spiritual realm, and how science describes mechanics. The manifestation of thought in the spiritual realm is very different than in the physical realm, which is most closely identified as "brain activity."

■ *It's interesting, though, if you think about it. If your brain says, "Lift your arm," you lift your arm. If you believe as much as you've been trained that wanting to lift your arm will make it happen, then you could probably do the same with something extraneous to the body. This is, of course, what we describe as "telekinesis."*

Absolutely, and I really think that a lot of the changes that we're seeing in our culture right now are preparing us to shift into a whole other level of consciousness. I recently wrote an article about the relatively new trend of text messaging, which we use in our cell phones and electronics. We are developing a whole system of symbols and abbreviations to replace language that is really an ancient form of communication! Way back in antiquity, a lot of the first writings were pictorial, where an entire concept, an entire idea would be expressed through a symbol, instead of writing it out through written language. Instead of a string of words, you can use one symbol that portrays the whole concept. That's part of going back to an ancient way of thinking: thinking in concepts, rather than words and linear thinking.

This is thinking "whole-istically." More people are connecting, because of this new tool. If you have a cell phone, you're connected. You're connected at all times, instantly and almost everywhere, to most any place in the world. I think about folks who talk about how the indigo children come in "connected" to a spiritual network. They're always in constant contact with each other. Culturally, we're moving into a position where, when these children come in and say, "I'm constantly connected," it won't be so weird, because the mass populace is now used to being constantly connected.

■ *So, in other words, you think that the matrix of communication networks that we've created is a positive thing?*

Absolutely. I think we're just moving to where that is now culturally considered "normal," so that when people can move things with their minds, without touching them and without being hooked up to technology, it won't be so strange for us. No doubt, by that time we will be accustomed to seeing people move things simply through brain activity. I think that, through technology, we're developing a way for the masses to encounter these ideas, these philosophical, perception-changing, paradigm-shifting ideas, before they actually come to the realization that people can manifest this without the technology.

■ *How do you see us, as we career into this vortex that we call the "Ascension," or the "Great Shift"? So much attention is focused around 2012. What is your perspective on how humankind is evolving?*

I think the very first thing that we have to pay attention to is our thoughts, mind to mind. Robert Monroe was one of the first people to put out books that became popular on "out-of-body experiences" and to research that holistic science.

■ *That's the director of the remote viewing organization.*

Yes, they do teach that there (they experiment with everything that can be defined "paranormal"), but he is most famous for facilitating the "out-of-body experience." What he called an "H-Band" of noise pollution around the Earth, he claimed was the product of uncontrolled thought. Annie Besant wrote a really wonderful book years ago about thought and what thoughts were. What she was saying about the energies of thoughts and how they manifest was that we really need to get control of our thoughts, if we are going to be able to move forward. We need to take personal responsibility for the content of our thought energy and get the noise level down: to get the mental pollution level down. Many people are talking about how exciting it will be, when we are linked telepathically, but you know what? At this moment on Planet Earth, I sure don't want to be! There's just too much noise pollution.

■ *I agree with that.*

I really don't want to connect with most people in that way; I would want some kind of filtration system applied to it. But if we take thought energy seriously (and that it is the first step in manifesting our reality), then we have to be a lot more careful about how we think of the world and of ourselves, through our internal dialogue and the energies that we send toward another person or situation. I think this really is the first step to our being able to make a grand leap forward.

■ *Indeed. In fact, you talk about this in the book: how group thought changes the antenna pattern. I found that very intriguing as well.*

Absolutely. Science has done study after study. The Princeton Engineering Anomalies Research (PEAR) program ran twelve years of study showing this, documenting over and over again that it does work. When people put themselves in groups to meditate and focus their group intention for a certain outcome, things changed, like bringing down the crime level in Washington. When people were sitting together, thinking about just that, the crime level fell; when they stopped, the crime level went right back to normal. These studies are documented, so it does work. And, as you know, the intention of healing through things like prayer, meditation, and those higher intentions are very powerful. . . .

■ *Most definitely. And of course now I'm going off the track a bit, but I'm sure you're familiar with the crop circle phenomena, which is one of my favorite discussions. I am always curious about people's experiences and their opinions about the phenomenon. Many people have talked about groups getting together and visualizing, focusing on a form or symbol, that then comes down on the crop.*

I can see where you can manipulate those energies. Unfortunately, there is so much controversy around the hoaxes of crop circles and I'm really sorry for that—because there is something to it. Many people dismiss it, thinking the formations are hoaxes, but they're not. I think people forget that there are earth energies that can be manipulated and interacted with. It doesn't necessarily have to be tied with ET or UFO activity.

No, I'm not talking about the UFO aspect of it. I'm referring to the reports of people who sit together, meditating on a specific formation or geometric form, and set their intention to create or manifest a pattern in the field.

They can; we can manipulate earth energy in such a way to affect the face of the Earth.

I'm a little uncomfortable with the word "manipulate," although I know you're coming from the scientific perspective. Let's say "to amplify and resonate" with the earth energies.

I use the word "manipulate" because it is our intention that we are specifically guiding toward one purpose. I can come into harmony with a tree, but I don't change the tree. If I want to make a change then I have to be very specific about what change I want.

. . . Focused intent.

Yes, you're right—that's much better than "manipulation." The power is in being able to interact with all sorts of energies that surround us in order to manifest change. I think it's quite wonderful when people can see something become manifest before their eyes, but that's not the point, you know.

That's exactly right. That's not the point. I am concerned about people getting hung up on the "phenomenon" aspect of spirit work. There is so much fascination now with the appearance of "orbs." It's nice to see these things, but let's not move our focus to photographing phenomena, which distracts us from being in the power of that inner explosion of energy.

Let's talk about other types of mind-body measurements. We were back talking about the measurements of waves of sound and light. And what else comes from the body?

Technology needed some time to catch up to what intuitives have been saying all along. And we just didn't have the technology to be able to take some of these measurements. At one time, there was no technology to be able to create or measure a radio wave either, but of course now we take that kind of thing for granted. Invisible light also carries information, but that was just science fiction at one time.

What healers have been saying is that they are sending specific energy out of their body toward another person. And now we have technology that can measure that, which is so sensitive that you have to be in a specially shielded room, because things like radio waves can interfere with it.

What they've been able to measure is the energy emanations that come from the hands of healers, and so they are finding it. One of the most fascinating things that they're finding, especially amongst healers, is that these energies are in direct phase alignment with the Schumann resonance, which is the same wave that transverses the Earth at 7.81 hertz per second. This is an energy wave that bounces between the cavity of the Earth's plane, the ground, and the top of the sky.

This energy wave is energized by lightning, and it creates a wave of light that travels around the Earth. What they've been able to find is that the energy that comes off the hands of healers is in direct correlation to that, in some manner. That was a very interesting find! What they've also found from healers is that they're almost like scanners. The energy that comes from them goes up and down the light scale, until it finds a frequency of interest and then it stays there, wherever the healer thinks they need to focus in on. They stay there for a moment and identify some very specific energy that they are connecting with, and then they go back to scanning again.

They have been able to measure these things in a very scientific way, and they have proven that these energies are emanating from us—energy at a higher frequency—that we weren't able to measure before. They're very, very faint, which is one of the reasons we couldn't measure them. It's like taking an EKG, of the heart or the brain, which measures electrical signals that are not faint at all: you can measure these things from across the room. But when you want to take magnetic energy readings of the heart and the brain, you have to be right on top of the person to be able to do it, because they're very faint in comparison. These subtler energies are even fainter still, and you have to be in a special shield to do them, but they have been documented.

■ *And what kind of equipment are they using?*

It's called a SQUID. It's a super-conductor; it goes into quantum physics. It's looking almost at the quantum level of what's going on. That is an acronym for "Superconducting Quantum Interface Device." It's pretty extraordinary.

■ *That is extraordinary. Now, what do you think of Kirlian photography? It became quite popular years ago at New Age fairs and in spiritual circles. Is there validity in this kind of investigation?*

I think it really helps to point us toward the idea of the energy sheath that's around the physical body. No matter what it is that they're taking (and there's a lot of controversy about what is being photographed), it doesn't matter. Honestly, to me it doesn't matter. What we can do with it is to see that there are energy templates that are invisible to us in the physical realm that we can measure, just like radio waves are invisible to us in the physical realm. We have to have very special equipment to measure that level of light. We can start working with these energy layers that help move science into a way of being able to quantify them, and energy workers can start dialoguing with science, because they have a way to say, "We can start quantifying these things and talking about them."

Barbara Brennan has really done some incredible work with that. She used to work for NASA and she has brought her scientific methodology into working with the "human energy field," as she calls it. Most people see this subtle energy body in layers, like an onion. She doesn't see it like that. She sees it as one thing, like a whole onion, and she sees different vibrations of that—different textures in them—and each one of those bodies has a different function that brings in different information for us to process.

■ *I did not know that Barbara Brennan had worked for NASA!*

Oh, absolutely. That's why I really pulled up her work quite a bit in *The Sage Age,* because of her methodology. What she's saying, and one of the reasons that I highlighted her in the book, is that for the most part, everything that intuitive practitioners do is personal and subjective. What Barbara Brennan is trying to do is to make it so that it's more quantifiable, so it's the same for everyone. She explains that, in the case of medical students, when they see a diseased part, they all see the same malfunction. What she wants to do with energy workers is when they're looking at auras, when they're looking at the human energy field or the subtle energy body, they should all see the same thing. She wants to make it less personal, less subjective, and bring it into something that science can quantify, so that we can study it to start making some use of it.

If we can start healing at that level and let the physical body just fall in line, once those energies are squared out, we will get rid of a lot of the problems with the health industry that are occurring today.

■ *Hmmm. I don't know I agree with the idea of quantifiable healing—as a psychic healer.*

As for the health industry, it sure seems that, as it stands at the moment, it doesn't have wellness as its primary objective!

No, it doesn't and neither do people. I think the one thing that folks forget (and I hear this a lot amongst intuitive practitioners in the healing arts) is the "pointing fingers" thing. One of the first things that's got to go is that we've got to quit pointing fingers at the other side of the bridge and saying they don't know what we're talking about ... or even worse: that they don't even know what **they** are talking about. We have to be able to both value all ways of knowing, and to find a way to work together to stop this pointing-finger business.

I think a lot of people don't understand the health industry is a consumer-driven market. Back in the day when X-rays first came out, people were so fascinated by these things that even if they had only a head cold, they wanted an X-ray. They demanded it. The doctors who would give it to them got the business. Until we found out that X-rays are actually harmful and the government started to regulate them, it was up to the doctor to decide whether it was worthwhile to take one or not.

It is a totally consumer-driven market. When people lose their fascination with illness and the need for illness, the entire industry will go away. But it's in that order: we don't get rid of the industry until we get rid of the need or desire for it.

■ *That's absolutely true. People can get into that trap even with holistic healing. I've been a healer for thirty years. People would come and lie down on that bed and basically say, "Fix me!" Then, they would feel great and go away, back to their inharmonious lifestyles. They would leave asking when they could make their next appointment. I tried to make them understand that if they thought the solution to their ailments was simply coming in to get "fixed" once a week, they weren't really taking responsibility for their own well-being; they weren't focused on being well.*

My understanding of the purpose of medicine in general is that it serves to break the cycle, to interrupt the cycle long enough for that person to realign with the highest templates that they can access for themselves, and the one that's the healthiest for them. Medicine is just to interrupt whatever current cycle they have that's bringing in or allowing the dis-ease. Medicine is not meant to cover it up or to fix it; it's simply to break the cycle for a moment and let that person remember health. The whole idea that most people have now is "fix me!"

Nobody from the outside can fix you.

No pill can fix you. No herb can fix you. No person outside of you can fix you. All they can do is disrupt the cycle that you're in, for the moment, and help give your body the time and space to be able to **remember health.** Because you know what perfect health is, you just need to find out what you need or what you're using this illness or this disease for. You have to get past that, for the illness or disease to go away. So there is a value to illness. Back in antiquity, if somebody in the tribe got sick, they were banished, because they didn't want everybody else in the tribe to get it. So, getting ill was not something anybody looked forward to or wanted ... but nowadays that's very different. Disease and illness bring attention to things that we want to pay attention to. As our evolution has grown and we can live in a state of luxury, if not banished from the tribe, we can use it for different things.

I think that's one of the reasons that disease persists: it still has value to a lot of people. When you look at an athlete and a diabetic, they're two sides of the same coin. They both have to pay attention to their bodies, at a level that most of us take for granted. They have to watch diet. They have to watch exercise. They have to watch themselves and pay attention to their body on an hour to hour basis. Some people have a preference for one side of the coin or the other, but they both have value. When I look at healers and intuitive practitioners, I see that they have learned something and they're passing that information on: I have been healed and I can show you how to do this.

That is very valuable information.

When we get rid of disease, guess what? All the healers are gone too. And some people are using that status: "I am special; I have conquered something; I am gifted. . . ." There is a lot of ego that goes with it. All of that has to go.

■ *Yes, all of it is going to go, for sure. I do tell people they won't be healing much longer, because it is time for individuals to be doing the healing themselves, or even better: forgetting about healing altogether . . . and just being well. I get a lot of reaction to that, because there are a lot of healers out there and they're looking at me like: "You're telling me that what I'm doing is not going to have any value anymore?"*

The best answer I give them is that it's just going to evolve into a higher form of compassionate support.

Yes. Absolutely. That is really, truly one of the reasons that disease persists. It is a system that we have come to rely on in some manner and when that system goes, part of our identity goes with it; who we think we are goes with that. If you have a disease, you start seeing your life and identity in connection with that disease, or if you're a healer you start seeing your life and identity in connection with that as well. That "personal identity" does have to go, because we are more than that. Reconnecting with ourselves at a higher level has very little to do with this personal identity.

A very dear friend, Juliet Nightingale, has had multiple near-death experiences. I listen to her, because she counsels people about their near-death experiences, trying to help them make sense of these things. She is probably the least connected person to this earth plane that I have access to. She is much more connected to herself at a soul level—not at the personality level. That's really what we're moving toward: giving up this personality level and detaching from it. To detach from what the personality thinks is so important and to go after those things that feed our soul—this is the reason why we're here.

■ *And in doing that, we will really be able to know the wonder of merging into the All, which we know is the purpose for why we're here.*

Yes. I think your earlier question was "How do I see us in the future?" Taking personal responsibility is the number one thing that lies before us: taking responsible for our thoughts, our well-being, our actions: doing that as our own process, rather than putting it out there, or projecting, in any way, that our well-being is dependent on others. They may add to it, as they help us explore and bring things to the forefront of our awareness, but it is nobody else's job but our own to do what we need to do for ourselves.

■ *Spoken like a true shamanka!*

It is essential to be truly accountable to yourself, to not need those outside rules—to not be accountable to an organization or to others—but to be accountable to yourself, and to live with integrity to the best of your ability.

■ *Let's talk a little bit more about sound and light, because we've gone off on the question of health, which is so important. But the theme about light and sound is so integral to your book. Can we talk about what we can learn from having much more awareness and understanding of the physical and the scientific and metaphysical study of light itself?*

One of the things that I'm working on now is a classification of sound, for energy workers to be able to understand the acoustics of sound. It goes all the way back to understanding how a circle works. Not what a circle **is,** but how a circle **works.** That has everything to do with understanding sound. It takes us into ancient geometry.

We are surrounded by symbols that are bringing through whole ideas for us that we sort of ignore, because for the most part, we've forgotten what they are. We need to go back and understand that, and see how they manifest into the world. And one of the ways that we can take geometry and manifest it into the world is to translate those symbols into frequencies of sound, because sounds are shape: physical shape.

If you throw an invisible box through the air, it moves the air, displacing the air particles as it travels. That is kind of what sound is like, when it moves through the air. When you encounter it in your physical being, you're literally having the physical shape of the sound penetrate your body. Sound healers all over the world are perfecting this, by understanding acoustics . . . by how a circle works.

Understanding the physical shape and how our body, as an antenna, resonates with it, and being able to use sounds for healing is very complementary to energy workers, who work strictly in the light realms, when they're working with the subtle energy body. I think these are very complementary studies and I think that if one person uses one . . . then they're intuitively positioned to work with others.

■ *Definitely.*

From the light spectrum perspective, I really don't know if there is any scientific data that would help anybody with that. In my personal opinion. I know we use the word "light" for a lot of things, and I think this is one place where science and intuitive art will never meet. The reason I say that is that I really don't think "light," as science understands it (as a by-product of heat), is the kind of Light intuitives work with. I think they're very different things. I think intuitives use the word "light," because that's the only thing in this realm that even comes close to simulating what it is that they're dealing with.

■ *But we do also work with rays. As a color healer, I've worked with colored light to heal different aspects of the body. So we do have some physical approaches. I'm not just only talking about the divine light, but some physical attempt to utilize light in a physical way.*

I think it's very helpful. Of course, quantum physics is all about light. Just that, for my own shamanic kind of work, it's not the same thing, as you know. Intuitives talk about "higher frequencies," to describe different energy realms. Well, science can measure all the way up at least to gamma rays and we can't get anywhere near those! They would destroy everything physical. Even X-rays, in high doses, will obliterate a physical constitution. They will kill it.

Scientifically, we know about all kinds of rays that would destroy matter, whereas you can be in the presence of divine light that is way over these other rays in the force of energy, and yet it doesn't harm anything. So that is another reason I'm saying they are not the same thing. Our investigations into the physical properties of light, through quantum physics, will never, ever reveal what intuitive art practitioners are dealing with. They can't. What they **can** measure in science is the shadow of that activity, just like a thought. As I said earlier, you will never be able to measure a thought, but we can measure the brain activity associated with the thought, so we can measure the physical effect of divine light entering into the material realm or intersecting with it, but we can't measure that divine light.

■ *...We can't measure "illumination."*

That's it. It can't ever be done. And as long as we continue to use the same word on both counts, scientists are going to continue to look at the only thing that they can look at, which is the spectrum of light ... and they're never going to find it. So they're going to go around and say it doesn't exist: "What you are talking about doesn't exist, because we're measuring it all and we can't find it."

Intuitives don't know the science enough to say: "I am using the word 'light,' but it's not that...."

It is the same for the word "dimension." In mathematics, a dimension is a form of measure, like length or height. To intuitives, it refers to realms of existence.

■ *That's right.*

So when you talk about "String Theory" and have to conceive a minimum of ten dimensions, intuitives run off and say: "Look! They are talking about ten dimensions of reality." But that is not what the scientists are talking about. They're talking about ten dimensions of length or height or a plane ... measures.

■ *On the other hand, there are master physicists, like Dr. Michio Kaku, who are talking about different dimensional realities.*

Well, okay, yes. He's the master string theorist, but when he talks about different planes of existence, he is talking about the multiverse, not String Theory, not the dimensions in String Theory, and that's where people get it confused. When he talks about ten or eleven dimensions of String Theory, he is talking about planes of measure.

When he is talking about other alternate realities, he's talking about parallel universes, the multiverse, the mini-world theory... and those things have nothing to do with each other.

■ *Isn't it so exciting, MaAnna, that this is what we're talking about now? It's just a wonderful, wonderful time to be alive.*

Yes, it is ... for all the changes that we've got coming, but some of them are going to be not only mind-boggling, but ethics-shattering. We really need to get a hold of ourselves about the "mind candy." There's a lot of material out there that discusses all of these ideas, but it does it in a novelty sort of way. It is just mind candy. People know these words, but they don't know what the things mean.

■ *Well, we're going to know pretty soon, because things are really starting to shift. I suggest to people I encounter in my workshops to be prepared to see a dinosaur appear at your dining room table pretty soon, as the dimensions merge!*

Right, we're even moving close to being able to combine technology with living tissue, in a way that is going to redefine what it is to be human. And if we can do things like perpetuate life to the point the people live to be three hundred years old, what will that mean to the resources on the planet?

■ *That does not sound like a good idea to me! What is so important is that people understand that they are immortal, without necessarily being physically alive, as we understand it.*

Yes, of course, but there are a lot of people pursuing this prolonged life span and there are just as many people who, with enough money, will be able to purchase it. That is really nothing, compared to what is on the horizon in the next ten to fifteen years. It's not far off at all. That is one of the reasons it was important to me to get clarity on some of these ideas, to help me help people understand what we're talking about and what we are being presented with, so that we can make responsible choices about science and technology and how we want to pursue them.

■ *And you've done a really masterful work with your book* The Sage Age. *My congratulations to you, for taking us all that much closer to merging science and spirit.*

■ ■ ■ ■ ■

MaAnna Stephenson's book is *The Sage Age:*
   Blending Science with Intuitive Wisdom

Her Web site is:
   www.sageage.net

Savikalpa Samadhi

■ ■ ■ ■ ■

Dr. Edgar Mitchell

On January 31, 1971, Navy Captain Dr. Edgar Mitchell embarked on a journey into outer space that resulted in his becoming the sixth man to walk on the moon. The Apollo 14 mission was NASA's third manned lunar landing. This historic journey ended safely nine days later on February 9, 1971. It was an audacious time in the history of mankind. For Mitchell, however, the most extraordinary journey was yet to come....

■ ■ ■ ■ ■

■ *Dr. Mitchell, welcome to* Beyond the Matrix. *I just can't imagine that I am actually speaking to someone who has walked on our mystical moon! I am going to open up right away with the obvious question: What in the world (or "beyond" the world) does it feel like to leave our beautiful planet and enter space, to then end up walking on the lunar surface?*

What more can an explorer ask than to be one of the first to go to another planet from Earth and to look around, to gather data, to explore, and then come back and tell everybody about it? What more can any explorer want? That's what it feels like!

■ *It probably isn't as far as you could go, but it certainly is one of the record-breaking journeys for any explorer, as you say.*

Hopefully, in due course, we will go further, but right now that is as far as we can go out there.

■ *... But it looks like we are getting ready to send men to Mars, doesn't it?*

In due course, I am sure we will, provided we aren't too stupid to blow ourselves up in the process.

■ *What do you think about that? Are we close, as so many believe, to the Armageddon scenario?*

As long as we don't quite recognize that the only thing that violence and warfare bring is more violence and warfare then we are in jeopardy, of course, of using the "ultimate" weapons. That would kind of be the end of things ... if we were stupid enough to do that.

■ *I have this deep sense that humanity or some kind of higher mind will reign and we will find our way back out of the turmoil.*

That is exactly what I hope and I do believe that, but I don't think it is assured. We have to choose, and certainly our choice has to be a little saner than some of the ones we've made in recent years.

■ *Can you tell us about some of your insights into the human experience and your personal impressions that support the idea that humankind is moving toward a New Dawn?*

As a matter of fact, every so often in human civilization—maybe every few hundred years, or even shorter periods—we go through a major transition, such as the fall of the Roman Empire, the Enlightenment, the Renaissance, the Industrial Revolution, and it appears that we are about ready for another such major change in the way we see reality and the way in which we see ourselves.

■ *I do think that the crises that we are facing today are forcing us to seek alternatives and that is a very positive aspect of these hurdles. For example, with the petrol crisis people are finally getting serious about finding ecological alternatives.*

A little late, but yes.

■ *It does seem that we have to go to the absolute breaking point before we fire ourselves up.*

One of my first public speeches after coming back from the moon was at Rotary International in Lucerne, Switzerland, about thirty-five years ago. One of the things that I mentioned was that we did have finite petroleum fossil fuel reserves and that we had better start thinking about it. Well, that was thirty-five years ago and right now we are getting close to what we'd have to call "peak" oil, whatever that really is.

■ *. . . But, on the other hand, we do hear that there are so many other potential sources. We just heard recently that there is a huge reserve outside of Brazil.*

Yes, there are reserves, but nevertheless they are still finite and, as we both agreed a moment ago, it is high time we started looking at alternatives.

■ *Indeed, especially as we have the sun, and as much power as we need we can probably harness from the sun.*

Yes, exactly, we can harness power from the sun and other sources. There are a lot of resources, but we need to get serious about looking at them. We have been a little bit negligent and I am afraid that the petroleum companies have been a little resistant.

■ *I would say "a little resistant" is a good way to describe it!*

Let's talk about your phenomenal journey. I still can't believe I am talking to someone who has walked on the moon—can you tell us what happened on that flight to the moon that resulted in your conclusions about the connection between consciousness and the Universe?

Well, first, let's just kind of set it up a little bit. Our flight, the third flight landing on the moon, was the first flight to begin to do science. The first flight, Apollo 11, was just to show we could do it and get off safely; the second, Apollo 12, was to show that we could pick our landing site and land there precisely when we wanted to. Each flight built upon the successes of the past. Our flight left following Apollo 13 (which didn't make it) and our mission was to start doing scientific research and geological experiments on the moon to find out what it really was all about.

After successfully completing all of that work and getting a chance to relax a little bit when we were heading home, the spacecraft was rotating in such a way that every two minutes in the window was a picture of the Earth, the moon, and the sun and a three hundred sixty degree panorama of the heavens. That, in itself, is truly awesome and it is just mind-boggling, because the stars and the heavens are ten times as bright in space, since there is no atmosphere like we have on Earth. Even on the highest mountains, it is ten times as bright in space with the number of stars and the brightness of the stars. Coming home, I realized, as I was looking out the window and watching this magnificent panorama, that the molecules in my body, and the molecules in the spacecraft and the molecules in my partners' bodies were prototyped and perhaps manufactured in an ancient generation of stars, because star systems are what create matter in our universe. Somehow, it suddenly became a very personal and visceral experience, rather than an intellectual one. I could feel it; it was inside me. It was accompanied by the ecstasy of bliss, which I had never, ever experienced before.

It turns out that people have been having these types of experiences for a long, long time but, as I had never studied esoteric experiences, I wasn't aware of it. This experience continued all the way home. We still had experiments to do, but the major part, since the moon landing, had been completed.

I realized that forever humans had been asking questions: "Who are we? How did we get here? Where are we going? What is our relationship to the Cosmos? What is it really all about?"

I realized that the story of humanity, as told by science (a relative newcomer in our history), was probably incomplete and flawed. I recognized how our cultural cosmology came generally out of religious and esoteric insights, which were archaic and perhaps wrong, and that now that we were a space-faring civilization (the first generation of space settlers), maybe we needed to ask those questions all over again. Maybe we needed to see them from a distant point of view, where we were observing the Earth from afar, witnessing the magnificence of this little planet and our solar system from quite a different perspective.

That was part of the insight that occurred.

When I got home—and this was so powerful that I couldn't turn loose of it—I started digging through the literature of science and found nothing that would satisfy me, with regard to what this experience was about. In the religious literature that I was familiar with, I found nothing, so I appealed to some scientists from Rice University, which is close to Houston, and I said: "Help me! Let's dig through the literature—ancient literature—and see if we can come up with something about this type of experience."

Soon after, they came back to me and said that what I had experienced sounded similar to what has been described in the Sanskrit of the ancient literature is an experience called "*Savikalpa Samadhi,*" which in Sanskrit means that you see things as they are in their individuality, but you experience them viscerally and emotionally— as a unity. That described perfectly what I was feeling and from it, I suddenly realized, from the ancient traditional mystical literature, that people had been having these types of experiences throughout history. I started digging more then and over the next few years, I got in touch with spiritual teachers and leaders from many cultures and all traditions, getting to the esoteric core of our common experience.

I realized that these types of transformative experiences are found in the literature and in the lore of virtually every culture on Earth. The experiences are essentially the same, but when we start to talk about them they become culturally embedded. Religious interpretations are applied. Unfortunately, when we explore these aspects of consciousness through our cultural expressions, they are transformed, somehow, into religious doctrines and we start arguing and fighting over whose religions are right or whose explanations are the best. We then are looking at exoteric forms, rather than the subtler esoteric wisdom.

■ *The fact that you are a keen scientist and have made this bridge is, of course, very exciting, at a time when we do recognize that science and spirit are merging. From the point of view of a person who has had that kind of cognitive, logical training, to have a "transformational experience" is remarkable and very exciting to be able to share with others.*

Well, that, of course, is the reason that I formed my Institute of Noetic Sciences thirty-five years ago, which brings science to look at precisely these types of phenomena. For the four hundred years since René Descartes, science has avoided looking at the subject of consciousness, thanks to the Cartesian dualism that has basically prevailed since then. Body-mind physicality were considered different realms that didn't interact and, of course, we now know that is not true. We have now discovered quantum science, which shows that they do interact. We are finally able to utilize the tools of science to look at these issues.

■ *Please, can you tell us more about the Institute of Noetic Sciences that you founded? What exactly are the activities of this organization?*

When we started out, our purpose was to use the tools of science to look at abnormalities in our understanding of consciousness, and to make consciousness a proper study, per se. Of course, science operates by anomaly; if we find things that don't fit within an established framework, one we believe is the correct one, then we have to find a way to create a new framework—a new paradigm. I was working and interested in so-called "parapsychological phenomenon," although I don't like the term "para."

We now know that most of it is explainable in quantum science and so there is nothing "para" or abnormal about it; it is just the way we are put together. That has been resolved through much of this work and it is to help look at the so-called anomalies in that perceptual understanding and realize that the so-called "psychic stuff" isn't really paranormal at all, but quite easily explainable once we get into the realm of quantum physics. That, of course, has been a "no-no" also, for eighty years or so, since quantum science was organized in the early part of the twentieth century. Nevertheless, we, the scientific community, have avoided even touching the concept of consciousness, leaving that to religion and philosophy. But we can no longer do that. We have made enough discoveries that simply require us to use the tool of quantum physics to help us understand mind-consciousness and broach the subject of spirituality.

■ *The time is, indeed, now.*

Yes. There are a couple of metaphors or analogies I like to use to help get this across to people. We have demonstrated many times in the laboratory that the interconnection between people, particularly the so-called "twin effect" between lovers, parents and children, can go across a Faraday Cage barrier that blocks electromagnetic level signals. It cannot be screened out by electromagnetic screening devices. That is a property called "nonlocality"; it is a property of quantum entanglement. Such experiments have been done hundreds of times, not only on humans, but on plants and animals; this property of entanglement is the basis of quantum science.

■ *Can you explain to us a little bit more, in lay terms, what you mean by "quantum entanglement"?*

Back at the beginning of the earlier part of the twentieth century, the early greats (Einstein, Dirac, Planck, and Schrödinger, to name a few) realized that there were some different properties emerging, when they looked at matter at the atomic level. Quantum physics emerged from these observations. Entanglement is the idea that, if subatomic matter is ever in a process together, and then is somehow separated, it nonetheless retains its connection as "quantum entangled" or "quantum co-related," no matter where the subatomic particles go in space. If something happens to one, such as if it's captured or some operation is performed upon it, the other subatomic particles that were entangled with it will instantaneously respond—a thing Einstein referred to as "that spooky action at a distance!"

We still don't really understand the basic mechanics of it, yet it is indeed real and has been demonstrated over and over again. The properties of quantum entangled particles are coherent: for example, light that is entangled is coherent light. That means that its waveforms are bound up and the particles that are entangled maintain this property of nonlocality, which means that they are instantaneously in communication with each other. We have also thought for eighty years that this does just not simply pertain to our scale size, our real world—that it was only a subatomic property. But in recent years, experiments have been conducted that prove that this does not only pertain to the subatomic world.

It pertains to **everything**.

Recently, about fourteen years ago, a colleague in Germany, Dr. Walter Schempp, discovered something called the "quantum hologram," while he was experimenting with MRI devices to improve them. The quantum hologram means that we now understand that all matter emits and reabsorbs coherent energy continuously. From the time of quantum physicist Max Planck, we thought that these emissions were random: they didn't really carry much, if any, information. Instead, Schempp's work shows that they are quantum entangled and nonlocal, and they carry information about each physical object: whether it's a coffee cup sitting in front of me, the desk, the headphones I am listening to, or myself.

Everything has an ephemeral image called a "hologram."

The ancients had similar ideas about that, as they believed that all of nature retained its experience ... they called it the "Akashic Record." People who have had such mystical experiences will find that science and spirit merge here—and that the quantum hologram concept really does give insights into how the Akashic Record can exist.

■ *Dr. Michio Kaku talks about how we now can actually change a quantum by changing its vibration, which is very exciting, because we know that thought is vibration . . . so we are starting to see the potential for the so-called "mind over matter" expression, seeking form in reality.*

That has been demonstrated in the laboratory many times now and I personally have been involved in this form of experimentation. In my book, I elaborate upon such experiences, although I was not trained in that kind of work and didn't really know much about it until the last thirty-seven years working with Noetics. I have also had my own experiences of being healed, for example, by intentionality and so-named "psychic processes." It all is included in the field of "mind over matter" or the "interaction of mind and matter" concept.

■ *This is pretty much what you were talking about, when you said that you had discovered the connection between consciousness and the Universe when you were having these awakenings on your flight back.*

That is all entanglement and these are all entangled ideas here. There are many different facets of it, but it's all coming in and is fairly easily explainable as natural . . . and the science that comes out of quantum entanglement.

■ *Can you tell us a little bit about your experience of actually touching foot on the moon's surface and how this affected the way you perceive reality and your own life?*

Well, setting foot on the moon didn't do that too much; we were just too busy. When we were physically working on the moon, we had deliberately programmed ourselves, within our timeline, to be at one hundred and twenty percent of human capacity. This was vital, in case equipment broke and we had to discard it or deal with unexpected events. As it turns out, we didn't have equipment breakage on the surface, but we still had to perform at one hundred and twenty percent capacity, so that meant we were really going and pushing ourselves to the limit. We didn't have a lot of time to be thinking of looking at nature and enjoying the experience of just being there.

■ *Right, but surely when you stepped on to the moon you must have had some awesome emotional experience.*

Yes, the awesome thing was stepping out and saying: "Wow! Here we are, this is so wonderful and it's different but, okay, get back to work, guys. We haven't got any time to waste!"

■ *That's understandable. To say the least, it had to be a severely high stress work environment.*

Yes, but when children ask what's it like to walk on the moon, I tell them it's like walking on a trampoline and your mum puts a couple of snowsuits on you. It is a little hard to move with the pressure suit on, but it is like walking on trampolines, where you have a springy step, but you can hardly move your body around with any mobility.

■ *Okay, I am going to ask you a question now that no doubt you have been asked before. What do you say to the people who claim that we never landed?*

Oh, it is very simple to handle that one. Do you really think the Soviets would let us get away with faking it? The whole thing was about a race into space with the Soviet Union and we got there first. They were watching everything that we were doing. If we had been faking it, they would have cried very loud and long about that, so that's just nonsense.

■ *How do you think that myth developed?*

...From people wanting their fifteen minutes of fame and being ignorant of reality. The answer to that question is that the evidence is so overwhelming to anybody who has any sophistication.

■ *So, you do your work on the moon, you get back in your craft and you head back and you say that you have these "spiritual epiphanies" on the journey back. How did this manifest for you?*

It manifests as what I call "the big picture approach." Right now, if anybody cares to look, the new pictures coming from the Hubble Space Telescope are so overwhelming and so powerful that they allow us to see so deeply into the Cosmos—deeper than we have ever been able to do before. We had magnificent views and vision with the naked eye—just from the spacecraft in space—but we could also look through our little telescopes, which we used to align our navigation platforms. We could see a bit closer to some of what the Hubble images are like and it is so utterly mind-boggling! It is so life-changing, you can hardly explain it; people just have to really experience it to believe it. The closest we can come to it, from our earthly perspective, is looking carefully at some of the remarkable Hubble images that have been brought back. If you can get into a meditative state when looking at some of these amazing images, it can trigger similar states of awareness.

■ *You conducted ESP experiments on the flight home, right?*

Yes, I had been interested in that work and I got interested in Dr. J. B. Rhine's work in the mid-1960s, which had studied that a bit. It so happened that a couple of physician friends of mine, whom I had met when I was in the program (before I flew), were also interested in Rhine's work. Together, we cooked up this little scheme to have a private ESP experiment in space, just to see how the enormous distance would affect the properties of telepathy. One of our members kind of blabbed it to the press and that blew our cover, but the data that was collected was checked by Dr. Rhine himself and by a Dr. Osis in New York, another parapsychologist, and it was as good as anything that had been done in the laboratory. It showed us that space distance didn't really make any difference.

■ *... But did it have to do with mind-linking with people on the Earth?*

It had to do with an experiment using the so-called "Zener" symbols that have been used in laboratories for forty years or so, repeating those types of experiments and concentrating on one of five Zener symbols: the star, cross, wavy line, circle, and square. We did these for about ten seconds at a time, in groups of about twenty-five symbols at a time, and then four people on Earth were trying to write down what they thought the ordering of the symbols was. They were significantly accurate at getting the way I had ordered them, by means of the way I was thinking about them. Statistically, chance could have produced those results only in one out of thirteen thousand experiments.

■ *During your experience, did you encounter any other kinds of craft in outer space?*

No, none up to the Apollo Program and through the Skylab Program, and having talked with the astronauts involved, there were no unexplained sightings at all.

■ ... *But there have been recently, right?*

Yes, I think there have been recently. I haven't talked to those astronauts who have personally experienced or confirmed any of that, but there was nothing up through the Apollo Program, at least in my experience. There were things observed that turned out to be parts of spacecraft or objects that we knew had been tracked and, in due course, we were able to identify them.

■ *What is your feeling about life beyond Earth? Obviously, when you talk about the greatness of the Universe, it seems that you have no question in your mind that life does exist beyond the planet.*

Of course I do! We have been visited; I have no doubts about that. The evidence, my knowledge, my research are there, and I have been well briefed, in fact, by what I call the "old-timers" who were involved in the Roswell incident—both in military and intelligence circles. I grew up in Roswell, so I was well acquainted with that area. From what I have been given from the old-timers who were involved with the Roswell incident sixty years ago, **I know that was a real visitation** and there have been many since then. I don't have any doubts about the reality of that.

■ *You know that was real? Do you feel in any danger by pronouncing that?*

No.

■ *That's good to hear, because obviously it is very "present" information in our circles of the metaphysical and the UFO field.*

I am not under any secrecy oath in that matter.

■ *That's good. What are some of the other areas of research that you believe that science still has to explore?*

What is coming to the foreground is this modern work in "quantum holography." It is putting a whole new face on many of the things we have taken for granted and causing questions to arise about some of the science that we have had in the past: for example, the Big Bang. A lot of people now are questioning whether the Big Bang is the correct source of the way the Universe originated. Many scientists looking at this don't think it is the right answer. The work that has come out of this quantum holographic formulism that my colleague discovered would suggest an alternative way that that could happen: that matter comes up the "zero point field." It seems very likely that British astronomer Fred Hoyle's notion of the "Steady State Universe" looks like it could very well be the right one. But we have a way to go before we can really demonstrate that.

■ *I talked to Dr. Kaku about the bubbling up continually of new life, new thought, new energy in the Universe, and that perhaps it has always existed and always will ... and that these Big Bangs are nothing more than bubbles appearing in the self-perpetuating physical universe. What do you have to say to that?*

That is exactly what the formulas that we are looking at would suggest: a so-called "quantum foam" in deep space. Space is not empty: there is quantum foam; there is un-manifest energy; particles; anti-particles popping into existence and disappearing continuously.

■ *We are now truly entering 2012 and this seems to be creating a lot of anticipation—some fearful and some joyful. As you know, the Mayans said that this was the beginning of the Fifth Sun. What is your feeling about this time reference?*

I don't place much thought in the Mayan calendar as being something prophetic. I believe that, if there is any real problem coming up here, it is due to our own activities. It is my opinion that the problems occurring right now have little to do with Mayan calendars, but rather with our own blindness, as to how much of our activity here is to our own detriment. We are destroying our environment; we are over-populating; we are doing all these sorts of things that are creating exponential growth on Earth. Earth is a finite space and exponential growth of anything and everything that is associated with humans simply cannot continue indefinitely. Buckminster Fuller pointed out, at the beginning of the Space Age, that we are the crew of spaceship Earth, but that we are a crew in mutiny... and how can you run a spaceship with a mutinous crew?

■ *How do you foresee the resolution of this extreme over-population explosion?*

It is clear that we either learn to limit human reproduction or we must learn to live more simply, conserving our resources and using our technologies for the greater good, instead of the way we are now. Our self-aggrandizement and the acquisition of material things, rather than providing the road to happiness, are, instead, the road to destruction. At this time, our tools have been largely used for our destruction, not for our betterment, and we have got to learn to do better than that.

■ *And quickly, I might add.*

And quickly.

■ *What are your feelings about the computer era and the potential for dark and light use of the computer technology that we have?*

I speak of all technologies. Science has no morality in and of itself; it is how we use it that is either good or bad. We tend to use our genius and our creativity for our destruction, as much as for our benefit, and that is where we need to make great changes and transform our consciousness.

■ *I would imagine that, if a lot more of the power on this planet, and individuals as well, came back from the moon and saw what you saw, there would be a lot more love for this little blue planet.*

I think so. As a matter of fact, we have started an initiative and it came out of Frank White's book. It's called *The Overview Effect*, which is to help promote this notion of private space flight and what is going on with Virgin Air ... and with other sectors of our economy— with private entrepreneurs going into space. Bob Bigelow has launched some through the Soviet system. This whole notion of seeing Earth from space is transformational. For many years, we have said that, if our political leaders could have a summit meeting in space, we would see a totally different political and social system on Earth. We have all agreed on that ... but we haven't gotten them to listen.

■ *What would be your message to the leaders of the world?*

The most urgent thing is that we have got to get this environmental issue under control. We have to address the fact that, on a large scale, civilization is not on a sustainable path. We are spending all this time fighting each other over whose god is the best god and whose religion is the best religion, and killing each other ... when the real answer to all this is cooperation on a global scale. Otherwise, this civilization is not going to exist in another hundred years or so.

We are going to do ourselves right out of a planet.

There is a group who would like to do that, but the corporate leaders in the United States, at this moment, have been leading it in the wrong direction as they have for some time now.

■ *Well, as you said, there are "interests" at the highest corporate levels that don't really want peace on this planet. It's simply not "profitable."*

I understand that.

■ *You spoke a moment ago of your talks with children. Do you work with children? I really do think that the children are our hope for the future and I would like to know that you are helping them to understand the experiences that you've had, and how they can start to take measures for their own future and that of the planet.*

I do try to speak to children whenever I am in a given area. Quite often, I speak in schools, and many of my friends who are teachers are totally in agreement with the things we are talking about. They are trying to teach a level of science and history and understanding that will bring the younger generation into total awareness of the issues that are confronting us right now.

■ *... And they really need it, because they are so bombarded with so much. So many children are hopeless about the future; they need to know that the future is in their hands and that they can determine the outcome.*

That is true: it is doable, but we are going to have to get on it and do it.

■ *Rather quickly, I would suggest yet again.*

Yes, rather quickly.

■ *Is there any last inspirational message that you would like to leave with us?*

I think we have tried to be inspirational as we have talked here and shared our ideas, yet we've not pulled the punches. We are in trouble here, as a global civilization, and we have got to learn to pull together, or we are not going to exist. We are going to run out of natural resources; we are destroying our environment; our water supply is in danger. Yet, all of these issues can be solved, if we pull together and start focusing on correcting the problems ... instead of creating bigger ones.

■ *Thank you, Dr. Mitchell. Let us truly hope that through the crises we are living, at this moment in time, we will learn to raise our consciousness from materialism and self-gratification to the awareness that we are all in this together, sharing a beautiful little spot of blue in the ocean of the Universe.*

■ ■ ■ ■ ■

Read more about Dr. Edgar Mitchell's works and visions through his book *The Way of the Explorer* and other works available through his sites:

www.edmitchellapollo14.com
www.noetics.org
www.quantrek.org

The Extraterrestrial
Reality

■ ■ ■ ■ ■

Stanton T. Friedman

Stanton Friedman is one of the ultimate voices we have today that speaks to the questions haunting the UFO/ET field: abduction, alien presence, and the governments' determination to conceal the truth. He is a real firecracker—a spirited soul with an utterly explosive wit and style. He has such a wealth of knowledge and experience that one conversation barely even brushed the surface. I have interviewed him on three separate occasions, each an exciting journey into the hidden truth that now is being revealed—thanks to courageous explorers, like him, who are dedicated to bringing it out at last. This is a compilation of our conversations on everything from abduction, to alien contact and the government cover-up.

■ ■ ■ ■ ■

■ *What first attracted me to you, Stan, was the strange connection I had to your recent book, Captured! (co-authored with Kathleen Marden)— the story of Betty and Barney Hill's UFO abduction experience.*

Many years ago, I was in Boston on business, where I met Dr. Benjamin Simon—the psychiatrist who conducted the hypnosis therapy on both Betty and Barney, in order to help them overcome the trauma of their truly dramatic abduction experience. We talked many hours about the story and his experience as the therapist who worked with these clearly traumatized people.

I was even privileged to visit his studio, where he showed me the life-sized model of one of the aliens, replicated through Betty's visions.

This abduction case was the first such event, as I recall, to ever have been openly discussed in the media. The fact that you are talking about this is very exciting, particularly since the book describes many details of the ordeal, revealed through the hypnosis experience for Betty and Barney Hill.

Can you give us an overview of what this book brings to light, in the never-ending debate over alien activity on Earth?

It is a new approach to the story of Betty and Barney Hill. Everybody has heard of the story and an awful lot of people have read *The Interrupted Journey* by John G. Fuller, which was a world best seller back in 1966, but Fuller only dealt with part of the data that was available. What is different here is a couple of primary elements.

One is that Kathleen, who is Betty's niece, had all the tapes of the regressions, and she transcribed them all, giving a much more well-rounded picture of what went on in all those sessions ... not just one view of each question.

The psychiatrist, Dr. Simon, was clever in how he handled things. He often repeated almost the same question, but in a different context—to see how much the answers would change. What is particular about Dr. Simon is not that he knew anything about flying saucers, because he didn't, but that his specialty was extracting memories of traumatic experiences, from people whose lives had obviously been affected by them. He worked with literally thousands of World War II veterans who had serious post-traumatic stress syndrome, which back then they called "shell shock." He worked out a technique utilizing very deep hypnosis, carefully extracting what the person knew and felt at the time, and then induced amnesia after each session.

It is kind of amazing how sometimes people think it was only one session that was conducted upon the Hills. The truth is that Betty and Barney were each hypnotized separately for months of weekly sessions, amnesia being induced at the end of each session, so that neither one knew what the other was saying in the sessions, always conducted in a soundproofed room.

One of the things that the "noisy negativists" (as I call them when I am being polite) try to say is that all this UFO "abduction stuff" is just the hypnotist pushing onto the so-called "abductee" his view of what must have happened. Well, in the first place, that is nonsense: not that there aren't any cases like that—of course there are—but in Dr. Simon's case what is perfectly clear is that—to the contrary—if anything, he was trying to push Barney into the position of saying that he had just absorbed Betty's dreams, which took place not too long after the original experience.

■ *He was playing the devil's advocate, in a broad sense.*

Yes, he was and Barney wasn't buying any of it. Here we have a case of the hypnotist, a world-class psychiatrist, you understand, but a hypnotist who was, if anything, pushing people **away** from an abduction viewpoint entirely. So, that is one of the important things that Kathleen has done, a comparative analysis of what Betty said, of what Barney said, and of what Simon was trying to get Barney to say.

Another important element of the book deals with the star map work, because the earlier book had Betty's drawing, made as a post-hypnotic suggestion (from Dr. Simon), of what she says she saw, while she was asking the leader of this alien crew where he was from. In the understatement of the month, she said, "I know you are not from around here!"

He showed her what I can only describe as a "three-dimensional map" or model, probably a hologram, but we didn't have that word back then.

She looks at the hologram and asks, "Where are you on the map?"

The clever alien says, "Well, do you know where **you** are on the map?"

"No!" Betty replies.

"Then how can I tell you where I am from, if you don't know where you are from?" says the wise alien commander.

Betty didn't know anything about astronomy ... not surprisingly: she was a social worker; Barney was active in civil rights activities. This wasn't the scientific world and, contrary to the popular mythology about them, they didn't watch science fiction. Somebody tried to foist this off on me two nights ago with: "Oh, Barney just saw a science fiction program."

It is nonsense.

Dr. Simon begins working with these two very sensible people and he gives Betty a post-hypnotic suggestion to draw this map later on, if, and only if, she can remember it accurately, and she does that.

It is in the book and it remains a big mystery.

The leader explains to Betty: "You have got sixteen points which stand for stars; you've got heavy trade routes, light trade routes, and occasional expeditions."

The map is in the book, but how are you going to find out what it means? I mean, we've got a galaxy that has got two hundred billion stars in it—well, obviously it doesn't have all the stars in the galaxy, but even our local neighborhood has gone within fifty-five light-years back, then we've got about a thousand stars. Now, we think there're two thousand, because we can detect much smaller stars than we could detect back then, as instrumentation has greatly improved. But, even if there were only one thousand, only forty-six of them are "like stars" to the sun!

I had better stress that "like stars" means not too hot, not too cold, not too old, not too new, not too close to another star, not too variable in their energy output (alternately freezing and frying is not a good way for life to develop, apparently), and so what does it mean?

I got a call, a few years later, from the head of the Aerial Phenomena Research Organization, APRO (one of the two major UFO groups in the States back then), asking if maybe I could help a woman named Marjorie Fish, who was working on the Hills' star map. She had visited Betty and had come up with a unique approach.

She was a third-grade schoolteacher, but she was a member of MENSA, the high-IQ organization—certainly one of the most objective people I had ever met. She was building three-dimensional models of our local neighborhood, to see if she could find a three-dimensional pattern that matched the two-dimensional one that Betty had drawn. I met with her; examined some of her models; helped discuss her work with Dr. J. Allen Hynek at a planetarium in Chicago; helped at a mutual UFO network symposium in Akron, Ohio; and published the first article. I also instigated work by Terence Dickinson, editor of the very respectable *Astronomy* magazine. His article about her work—and he talked to a lot of people—verified that it was accurate and carefully done, and it got more response than anything the magazine had ever published before—or since.

I recently talked to Terry, and we both agreed it is remarkable how well Marjorie's work stood the test of time. Basically, what she did was to take beads and string them on a nylon cord in a big wooden frame. She would place the cord in the right two-dimensional location indicated in Betty's drawings and then look at it from all different directions, to see if she could find a three-dimensional pattern that matched the two-dimensional one Betty had drawn.

It was quite a remarkable piece of work; she built a total of twenty-six different three-dimensional models, but the problem that she had was getting good "distance" data. You see, to astronomers it doesn't matter where or how far away stars in the local neighborhood are to map the stars. You need two angles, you don't need the distance, and it doesn't matter if it is twenty light-years away or if it is thirty light-years away. You are still looking at the right star.

However, if you are going to build a 3D model, you need to know the distances. It was only after a new catalog of the distances of nearby stars came out, by Wilhelm Gliese, that Marjorie found one, and only one, pattern that matched. As you can imagine, it was a big day.

■ *Phew! I can certainly imagine the amazement when the maps aligned!*

She's not a very emotional person, but she was certainly pleased that day, let's put it that way. What this did was to enable us to identify all the stars in the pattern and it turns out, most remarkably (and this was not known before this work was done), that all the pattern stars are the right kind for planets and life. They are sun-like stars. That's sixteen and only five percent of the stars in that neighborhood are the "right kind," but all the right kind in that three-dimensional volume of space are part of a pattern and all the pattern stars are the "right kind." The chances of that being an accident, just coincidence, are one in ten thousand! Some say one in a million, but being a very conservative guy like me, I say one in ten thousand.

■ *I wouldn't have taken you for a "conservative" guy!*

Well, I am conservative; I'm a nuclear physicist who has studied a lot of evidence, and you won't find me sticking my neck out, until I am pretty sure of what I am saying. I have a great big gray basket. But in this case what is really exciting is that the base stars in this pattern—the two stars that were close to each other with very heavy trade routes between them—turned out to be Zeta[1] and Zeta[2] Reticuli (that is, the constellation of Reticulum). It means "the net" in Latin.

You can't see them from here. You have to go below the equator.

Now, consider this situation: the nearest other star to the sun is about 4.3 light-years away. These two stars, from the sun, are about 39.3 light-years—it's just down the street; I mean, the galaxy is a hundred thousand light-years across, so it's next door. More intriguingly, the two stars are only one-eighth of a light-year apart from each other, which means they are thirty-five times closer to each other than the sun is to the next star over.

We're out in the boondocks, but they've got next-door neighbors and, ah, one additional little point: those two stars are a cool billion years older than the sun! One would expect, using "Friedman's Law," that technological progress comes from doing things differently in an unpredictable way, and that the guys out there probably know some things that we don't know. Just look back a hundred years and see how much things have changed.

You don't need a billion; maybe they were only a thousand years ahead—that's good enough!

None of that work had appeared in the original book about Betty and Barney, and none of that really appeared in a book since then, so *Captured!* gives people a chance to look at that.

It also deals with the criticism of Marjorie's work. It is amazing how the people that attacked her work—every single one—misrepresented what she did, which I find quite remarkable. Friedman's rules (I made them up for debunkers): *What the public doesn't know,*

I'm not going to tell them. If you can't attack the data, attack the people: it's easier. Don't bother me with the facts; my mind is made up. And: Do your research by proclamation—investigation's too much trouble, you know; it might take some time and effort!

Captured! has done well; it has had excellent reviews. Kathleen was trained in sociology and education, sort of an unusual combination with a nuclear physicist, but she was also active with the Mutual UFO Network (MUFON). For ten years, she handled the certification of field investigators.

■ *Did Kathleen become active because of her aunt's story and experience?*

In a way. She knew about part of the experience the day after it happened, because Betty called Kathleen's mother (Betty's younger sister) the next day and Kathleen was there. She was a teenager at the time and so she heard about it then. They visited Betty and Barney not too long thereafter. Kathleen saw the strange shiny marks left on the car; there were a lot of things that she was well aware of, right from the beginning. Then there came a period of time when she was doubtful, because some of Betty's own attempts at what might be called "UFO investigation" didn't really match up to reality. She worked too hard at it, so to speak, and jumped to too many conclusions. The whole point of the exercise is that, with Kathleen having had almost total access to Betty the last five to ten years of her life, she became the executor of Betty's estate.

She still has all the papers.

She is preparing everything for the University of New Hampshire archives, since she had a real opportunity to talk about it with Betty. Betty was not a shy person; she expressed her opinions. She was in my movie *UFOs Are Real,* way back then, as was Marjorie Fish, and another interesting person, Dr. Walter Mitchell. He was an astronomer at Ohio State University who helped Marjorie in her work, and stated on camera how accurate he and his students found those three-dimensional models to be.

I hadn't fully appreciated this, because I had become interested before reading about the Betty and Barney Hill case. The first book I read about it was way back in 1958, and the publicity about the case was in the mid-1960s. But I hadn't realized until both Kathleen and I were signing books at the Mutual UFO Network Symposium in Denver, in early August, and there were so many people who said, "This is the book that first got me interested in UFOs." People had heard about the case but hadn't realized that there was a lot more to it than what they heard about when the first book was out.

■ *Not everybody really knows the story of Betty and Barney, so could we really do a quick overview of what occurred and what was recorded by these two people?*

It was September 19, 1961. They were coming south from a quick trip—a sort of a delayed honeymoon—coming down from Canada to their home in Portsmouth, New Hampshire, which was between Massachusetts and Maine, right on the Atlantic Coast. Coming down through the White Mountains, Betty spotted a UFO, and they watched it for some time. It was late at night, nobody around, and Barney was driving. Betty told him she saw something that looked like a UFO and he sort of rejected it, saying, "There aren't any UFOs! It must be an airplane!"

"Yeah, well, look at how it's moving!" Betty answered.

Barney insisted, "Well, maybe it's a helicopter. There aren't any UFOs!"

They talked about it and they actually stopped the car three times, while they looked at the thing, because it came close and it was behaving in a peculiar fashion—a star-step flight pattern—which made no sense, because planes can't do that! They heard some strange noises, too. Barney got out of the car, looking with binoculars, to try to figure out what they were seeing, and he was able to see windows on the craft with beings standing behind them! This round object was just a few hundred feet away, hovering in the sky, and he suddenly felt terrified, because he was getting a thought from the being onboard the craft.

He dashed back to the car and they took off and, to make a long story short, eventually they stopped. I have been to the actual location where this happened, practically out in the middle of nowhere (in Lincoln, New Hampshire), and it is clear that the only way Barney would have gone off the main road (onto a secondary road and then onto a tertiary road) was if somebody was controlling his actions. He would never have done that.

An important aspect to this whole story is that the aliens seemed to control Betty and Barney—this turns up in loads of abduction cases—without putting a gun to their head or a knife to their throat, you know, like the standard earthling approach to this sort of thing. Apparently, let's call it "telepathically," they were able to control them—I don't know what other term to use.

■ *We could probably describe it as "mind control."*

Yes, mind control and, incidentally—think about that for a minute—wouldn't every government love to be able to control the soldiers of their enemies, just commanding them mentally: "Put down your guns!"

■ *We might even say they already know how to do that ... even if it is still relatively covert.*

There are many reasons for a government cover-up, but certainly one is to be able to figure out the technology of the aliens, not just faster, more maneuverable craft and silent flight and all their other secrets, but this mind-to-mind kind of dominance, which is really scary.

They were taken onboard and, soon after, examined in separate rooms: stick in a little needle here; scrape a little skin there; clip a little hair there. There were some curious moments, like when the aliens came into the room where Betty was (they had her and Barney in separate rooms) and the "investigators" started pulling at her teeth. Betty asked what they were doing—what was it all about?

They were perplexed as to why Barney's teeth came out (he had dentures) and hers didn't!

■ *Betty seemed to have a good camaraderie with, let's say, the "commander." I think she asked if she could take back a book as evidence—what was it? I don't remember.*

Well, yes, she did. She told him she wanted proof.

■ *He was giving her something and the others in the crew made him stop doing that. Is that correct?*

Yes, she says that he asked, "What would you like?"

She grabbed what she later described as a "book." It wasn't in the English language or anything like that, but there seemed to be columns of symbols there. She had that and was holding on to it, to bring back. It was only a little while later, when they were about to be returned, that the leader told her that they had decided that she wouldn't remember what had happened and so, he took back the book . . . much to Betty's distress.

"I will, too, remember!" she told him. Well, she got the last laugh on that one, I guess. But we don't have a souvenir.

Now, on the original civilian investigation of the Roswell incident, I know we have pieces and parts, but abduction cases don't seem to have people bringing back souvenirs: no "his 'n hers" towels, no matchbooks, no T-shirts, no mugs! That is part of this picture of mind control. We take for granted that people are all different—sensitivity to drugs, for example; our genetics are all different, that sort of thing—but it may be that aliens still need to learn about why some people react one way and some react another way. It is not an easy thing to go kidnapping earthlings, examining them, and putting them back out, especially if you want to be kind to dumb animals, which is kind of the way I tend to describe how they treat us.

We sedate polar bears and attach little devices that send signals back to a satellite, which then sends signals back to a researcher, so that he can figure out where the polar bear is at all times. Now, what do you suppose that polar bear tells his friends when he goes home for lunch: "I saw these crazy looking characters and they did something weird to me."

■ *It is interesting that you would make that comment, because I have just given a workshop in Amsterdam and we were talking about how it will be when we have global contact . . . and the concept of dark and light aliens, in the end, is not much different than dark and light human beings. Of course, we don't really think too much about the torture that we put animals through to do our tests, and yet the idea that alien beings would study us in the same way is beyond horrific to us. I think that is a very big statement about human consciousness.*

Remember Copernicus' book, in which he foolishly expounded the notion that the Earth is not the center of the Universe: the sun is? Just one little step over—the book was banned by the Church for three hundred years. We then got to the point where we recognized that the sun wasn't the middle of the Universe either. Maybe it's the middle of the galaxy, but now we know **that** isn't true and our galaxy isn't the middle of the Universe.

There is another very important part to this. Back in the time of Copernicus, Bishop Ussher went through the Bible carefully (how many generations, who begat whom, and how old whoever was) and concluded that the Earth was created, I think it was, the 23rd of October, 4004 BC—roughly six thousand years ago. Now, with radioactive dating and a whole bunch of other physics-related information, we know that our planets may be four and a half billion years old, and the galaxy is at least thirteen billion years old!

Suddenly, we realize there could be advanced civilizations all over the place, about which we still know nothing!

That bothers some people; we want to be on top of the heap. It is part of the difference between ufology and the SETI (Search for Extra-Terrestrial Intelligence) movement—I call it "silly effort to investigate"!

We presume that there are aliens out there, who are sending us signals, and that we should be able to figure out just what kind of communication systems they have, and then interpret those signals, as if they are somehow trying to attract our attention, to tell us all the secrets of the Universe.

My view is that we know so little about what is going on outside our own place here that we are much more like the gorillas in a nature reserve in Africa, who know nothing at all about what is going on outside the reserve.

Our egos are in for a beating here.

Now part of what goes with that is the whole question of power. On my Web site, I have a paper called "The UFO WHY? Questions," in which I get into the question of why would governments cover this up, as they certainly are. You can't imagine the blacked out and whited out documents from the government that we get under the

Freedom of Information Act. You can read one line per page and the rest is whited out—very helpful!

People ask me, "Why would aliens come here? And why would they come this long way through these huge distances?" I don't consider thirty-nine light-years a huge distance, not when Andromeda is two million light-years out there! "Why would they come all this way and crash in New Mexico?"

Well, it's a foolish question, because every indication is that the little craft don't do the travel between the stars; the heavy lifting is done by the mother ships, which people are seeing. There have been a lot of good cases of reported craft being identified as a half a mile to a mile long. The direct analogy here, which might make that easier to understand, is that the United States Navy has nuclear powered aircraft carriers; they are huge—about twelve hundred feet long—and they can go without refueling for eighteen years.

■ *Eighteen years? You can't be serious.*

Indeed! I'm a nuclear guy and I take great pride in the way they've arranged that. They use burnable poisons—materials that absorb neutrons—and these get used up at about the same rate as the fuel gets used up, so it keeps grinding and grinding away. Compare that to seventy-five small airplanes; they are certainly small in comparison to the carrier, and they can go three hours without refueling! So, in other words, the small craft clearly aren't just coming here from Zeta Reticuli. They are passengers on a big space carrier.

The second question—"Why would they come here?"—in my "why" questions: I give about twenty reasons for them coming to Planet Earth. In our weekly broadcast, "idiocy in the boondocks," on the radio stations out there, it is the honeymoon castle, this corner of the galaxy, or "gas/food/lodging: next exit."

We sometimes forget how many different reasons there are for people to travel long distances, and especially Americans. You would think they would remember the 49ers! In 1849, thousands of people went west, really out in the middle of nowhere. Why? Because gold was discovered. Why did they go up to Alaska at the end of the nineteenth century? Gold was discovered there too!

It turns out that Earth is the densest planet in the solar system. I don't mean the people, although that is probably true too, but a cubic meter of Earth weighs more than a cubic meter of any other planet in the solar system. (You may say, "Who cares?")

That means that there are more heavy metals here. Not only those such as lead, which is only eleven times denser than water, but others—like tungsten, uranium, gold, rhenium, and osmium, which are close to nineteen to twenty times denser than water. Many of these metals have very special properties, which we didn't know a darned thing about in the recent past. Do you know what the major use for uranium was one hundred years ago? It was a glaze on yellow pottery. Alien visitors may be coming after stuff that we place no value on.

There are many reports of UFOs coming up out of the ocean. Maybe they are picking up nodules of pure metals that are down there. Who knows? Maybe they are mining asteroids!

■ *Can you speak about the reports of craft having been seen hovering over a lake or emptying out a lake or just sucking untold quantities of water?*

There are reports like that, not only of pulling in water—and we are the only planet in the solar system that is richly covered in water, a lot of it. I don't mean that there isn't water in other places; I'm just saying that if you look at the total picture, we've got a lot of oceans out there; three-quarters of the planet is covered with it and some of it is pretty deep—miles and miles.

There are reports of them sucking up water from a lake; that's freshwater.

There are also reports of them going beneath the ocean and coming back up out of it; we refer to these as unidentified submerged objects (USOs). These craft may work similarly to the electromagnetic submarine, which takes advantage of the fact that seawater—not freshwater—is a good electrical conductor. It turns out you can solve all the problems of high-speed flight if you ionize the air, creating an electrically conducting fluid.

It is like what happens when a meteor comes in. What you see isn't the meteor . . . it is the glow around it, because it has ionized the air . . . broken up the atoms. You can solve all the problems of high-speed flight, control lift drag, heating, sonic boom production, radar profile . . . all these things. Seawater is another electrically conducting fluid, so, that may be a part of the whole picture here.

When it comes to the question of why they come here at this particular time—we can surely realize that there have been signs of sightings throughout time: go back to the Bible and Dr. Barry Downing's book *The Bible and Flying Saucers*. He goes through the Old and New Testaments, and finds a lot of flying saucer events. They weren't called "saucers," perhaps, but they are described in those passages of religious rhetoric.

If you go back and think about it for a minute, I think that one major reason for their coming here is because we are threatening to be able to go out there, as horrible a thought as that must be to our alien neighbors. There were three events at the end of World War II that would have told any smart alien that soon this primitive earth society, whose major activity is tribal warfare—soon these guys are going to be bothering us ("soon" meaning one hundred years, which is nothing on a cosmic timescale).

The first of these three events was the detonation of nuclear weapons, which leave their signs in the atmosphere and a few places on the planet—this is right after the war now, I'm thinking, where things are very highly radioactive and very easily measured. The second was the use of rockets. Certainly everyone knows that the V-1 and V-2 rockets that were being fired from Germany to England were not being used to carry the mail. They were being used to **destroy.** The third was the development of electronics, sort of primitive, but radar, powerful radar, which was one of the great secrets of the Second World War. The Germans didn't know that the English had developed it.

I have to tell you a funny little story. Back in 1938, before the war started, a German general heard that the English were building these two-hundred-foot-tall towers, with cross members on them, and he figured: "Well, **we're** working on radar in secret ... so that must be what the Brits are doing. We've got to figure out what frequencies they are using, the characteristics of the radar, and so forth."

They sent the *Graf Zeppelin*—over seven hundred feet of it—flying along slowly, and these towers were right alongside of the English Channel, facing Europe. Loaded with all kinds of gear, they flew the *Zeppelin* along, at about seventy miles per hour, to listen for signals, but didn't find anything. Meanwhile, the Brits picked up on this craft's movements. The Germans tried it a second time and they still didn't hear anything, so they presumed that the Brits didn't have radar. As it turns out, the Brits **had** developed radar! They were just using a frequency that was ten times greater than what the Germans were using.

The scientists on both sides of the English Channel were certainly aware of each other: they published in the same journals; they went to some of the same schools. So, if the Germans couldn't figure out what the British would use (considering how close the civilizations are, not just geographically, but psychologically), how in the world would a SETI guy figure out what somebody else is going to be using out there in space?

At any rate, you put all those three things together—nuclear weapons, rockets, and powerful electronics—and there was only one place in the world in July 1947 where you could study all three of those together. That was southern New Mexico. That was where the first atom bomb was tested: Trinity Site, White Sands Missile Range. That's where we tracked the rockets, which unfortunately didn't always go where they were supposed to go. Some even went south to Mexico instead of north.

■ *And that is where Roswell is, right?*

Ah, you've got it, you've gotten there!

What I am suggesting is that it is perfectly natural for every society to be concerned about its own security. Doesn't everybody we know want that? I think that means you've got to keep tabs on the predators in the neighborhood, but only keep close tabs on those predators that show signs of being able to bother you.

A simple question: Who speaks for Planet Earth: the president of the United States? The Americans would say, "Why don't they land on the White House lawn and open dialogue with the president?" as if he speaks for Planet Earth—almost seven billion earthlings! There are three hundred million Americans and he has trouble speaking for all of **them** much of the time. He certainly doesn't speak for the whole planet.

Besides which, incidentally, there were loads of sightings over the White House back in 1952, the biggest year in Project Blue Book's history. Military pilots were issued orders in the summer of 1952 to shoot down UFOs if they didn't land when instructed to do so. As one Air Force general publicly stated, "We scrambled U.S. jets after UFOs hundreds of times."

We have proof of this. The head of the American Rocket Society, as a matter of fact, a guy named Farnsworth, sent a letter to President Truman, saying he didn't think that was a good policy . . . and I tend to agree with him. But what I am suggesting is that, when the aliens did try to land on the White House lawn, we weren't very friendly about it.

There is a great book out about this, *Shoot Them Down!* by Frank Feschino Jr., for which I wrote both the foreword and the epilogue.

Well, you know it is a strange world. We are all concerned about survival and security, and there isn't anybody who truly speaks for the planet. You know, we Americans (that side of me anyway) say that we believe in freedom and democracy and elections. However, when you try to figure out how we are going to select the person who will eventually speak to the galactic federation for Planet Earth, it certainly appears we won't be able to manage a global election, when we can't even handle them at the national levels.

■ *I would like to know your opinion as to the following: what do you think will happen when we have global contact—not just sightings, but absolute confirmation around the world? I really believe that contact is imminent. What would happen, in your opinion (and I am talking conjecture, although I know you don't like to do that, because you are a scientist), when people all over the world start seeing fleets of crafts flying over major cities? What do you think?*

Well, if we could just stop shooting at them! . . .

Again, I think what we need is someone who speaks for Planet Earth, just as we need to make it safe for aliens to land. They tried over Washington, D.C., in 1952, and we sent up airplanes to try to shoot them down! They were picked up on radar and there were many trips over a three-week period in July.

■ *Do you think they were trying to make contact?*

No, I think on that occasion, they were checking us out. Of course, we don't know if there has been contact on a secret level.

■ *Personally, I am curious to see if the contact information would come from the government or directly from the source. Just think, if you had fleets of manned craft of highly intelligent beings ... they would first of all be able to take over the media.*

That's right. I think if they wanted to communicate to us, "Hey, guys, stop acting like idiots!" (which would be a reasonable thing to say), I think they could. Between our radio and television broadcast systems, I am sure smart aliens could figure out how to send their signal in on those same frequencies.

We are putting out lots of information. We also have radar systems watching the skies and spy satellites looking down and radio systems looking up, so we know that they are coming here and we know what their capabilities are. Surely they know we know, because when you have been tracked on radar you know it; you can tell when the signal starts and stops.

■ *Many people are concerned that they may be anticipating a military response, but we are hoping that the human outreach can extend past that barrier.*

Remember, now, that they have abducted lots of people; they have examined a lot of specimens and studied human beings. Most of those people have not come forward, because their memories have been blacked out. It has been my experience that only ten percent of people who have had a sighting have reported it. I have checked all my audiences: ten percent have seen one, but only ten percent of that ten percent have reported what they have seen, because of the fear of ridicule.

If that applies to just a standard "run of the mill" UFO sighting, then how about an abduction case? How likely is it that people are going to want to come forward? Admittedly, the best abduction researchers have investigated well over one thousand cases, but I am sure that is just the tip of the iceberg. What I am saying is that they **have** picked up specimens. They can also watch our television, which is a scary prospect (especially if they watch the commercials), and try to figure out what is going on with humans.

■ *They will assume the whole population is on meds!*

What I am saying is that I don't think they are going to send fleets over our cities, because they know that we know that they are coming here; the governments of the planet know that. Maybe there has even been an agreement that they won't go public: they can pick up their specimens and check things out without being shot at anymore, as long as they don't disturb the system per se, like grabbing the president, for example. That would get messy.

■ *How do you think crop circles fit in to all of this? Have you been in any of the formations?*

I have been to one crop circle; I have looked at a lot of pretty pictures and frankly, my opinion is that I don't have an opinion as to what these mean or who is making them. They are interesting; some of them very exotic and complicated—a whole bunch of circles of differing diameters in a special pattern and all that. I don't know what's going on.

■ *I would like to think, for example, that these are communications from other intelligence, more "multidimensional" than "extraterrestrial." For some reason, I visualize geometric patterns in the sky (rather than squadrons in warlike formations) when these crafts start to arrive. Wouldn't it be amazing to see them offer us a light show, with geometric kaleidoscope-type effects? It sure would be a great way to communicate intelligence, wouldn't it?*

It would be spectacular!

As for the crop circles, we can think of all kinds of reasons for them. Maybe aliens from two different civilizations are competing, to see who can make the prettiest and the most complicated and the fastest! There may be a judge up there who says, "Okay, you guys win for that one; tomorrow we will meet again at another place."

■ *They do definitely seem to fit into a picture of contact from beyond, much different than the information we are getting from abductees.*

What I am saying is that these guys are doing their thing, examining humans. Anybody with any sense wants to pay attention to his own survival and security and, as I said, anybody studying us for any length of time knows we are a primitive society, whose major activity is tribal warfare. They don't seem to be nasty guys—we are!

Thirty thousand children died yesterday of preventable disease and starvation and again, today, another thirty thousand will, and tomorrow too, and yet we will spend a trillion dollars collectively, on the planet this year, on things military. Something is wrong with that picture.

■ *Not to mention the secret laboratories, where they keep creating more viruses to unleash on the public, to kill more people. Talk about a self-exterminating civilization!*

Hey, death is a common way of life around here! That is a terrible thing to say, but it's true. In other words, try to look at us from their viewpoint. They see that we don't care about life, so why should they?

■ *I would like to believe that, just as we have duality on this planet, so would any other life forms in the three-dimensional construct. There would also be polarity of dark and light, and I would like to think that even though there are these funky ETs, I do believe that there are also very light ones, who will be coming in as well. Not only the ones that are working for the government, or the Greys or the classic stereotypes that we've got, but a broad expanse of intelligence flooding in to participate in earth affairs, once the gate is open.*

In fact, referring back to the book *Captured!*, you get a certain amount of insight into how the aliens were behaving. They were trying to avoid giving pain. They were curious . . . they treated Betty and Barney with a certain amount of respect—I mean, they did their thing, but they didn't put a gun to their heads, or a knife to their backs, the way we do things down here.

■ *I found it fascinating that the commander was originally willing to give her the book . . . that certainly seems like a desire to make a cultural exchange. Too bad he didn't!*

I think it was out of generosity. They decided that she wouldn't remember what had happened, so why give her anything that might be more distressing on her return? They have probably handled other earthlings, who they found had much more difficulty with the experience, partly because the government hasn't told them there are aliens out there.

You know, if your kids wander out into the forest, you tell them there are wolves, but our government hasn't told us there are wolves

or good guys out there. One thing—I have another paper on my Web site, that is: "Government UFO Lies."

I know it may come as a shock, but there have been occasions when the government has lied about UFOs.

■ *You wouldn't fool me about a thing like that now, would you, Stan?*

I sure wouldn't . . . and I don't discriminate. I have lies from the CIA and the FBI and the Air Force and the NSA. They've all been lying all over the place! So, you have to ask: "If there's nothing to this subject, why are you doing all the lying about it?"

Some of those lies are even about me, and, frankly, I resent that.

I have got UFO documents (it took me five years to get them under the Freedom of Information Act), by CIA definition, mind you, where you can read little more than eight meaningless words to the page—everything else is blacked out.

So the question has to be to the CIA: "Okay, this is twenty-five years' old! Executive Order 12958 says that 'you have to show a reason why you are withholding this information.' It is all supposed to be automatically declassified after twenty-five years, unless you can establish a reason."

In other words, put them right into the glare of publicity, if you want. Most media people, in my acquaintance anyway, are totally unaware of this vast amount of information that has yet to be released. They buy into the Air Force nonsense that Blue Book was **it**.

In my book *Top Secret/MAJIC*, I have a five-page response to a letter from the Air Force congressional liaison to a senator Patty Murray from Washington State, giving her the latest scoop, the same as it was fifteen years before, on the Air Force position. It is so full of holes that I wrote a five-page letter! I think any journalist looking at that would say: "Hey, he's right: they are not able to provide a sensible response to a question from a U.S. senator. **They lied.**"

You know, these are strong words, but I have it spelled out on my site, where I list the lies from a number of government agencies.

■ *That's not necessarily good for your life expectancy, Stan!*

I know I come on very, very strong. I am not an apologist ufologist. I am not a closet ufologist. I tell it like it is.

■ *And we love you for it! But is confronting the CIA and other government organizations and calling them liars necessarily "life enhancing"?*

I think so—if you do it publicly. You just go after them and you show the evidence. Any journalist can do that, if they read the documentation.

■ *I don't mean to say that you are naive, because of course you know I respect you without question, but the idea that a journalist could "go after them if he wanted to" makes me a little bit curious.*

I think that we have to put them in the position of defending actions that they have already taken.

■ *The media seem more intent upon scaring us to death. They sure seem to want us to think the aliens are going to be wolves, not good guys! The majority of people who do believe in alien contact are more in the fear mentality than joyous anticipation.*

Yes, people need to start thinking about what it means to them, if the government has withheld something that important for over sixty years, and if aliens are coming here—both of which seem to be true—to do their thing...whatever that will be.

How about if the media start to push the government to grant amnesty for any former military person who wants to talk about a UFO experience more than fifty years ago?

■ *Let me interject something here. I recently got sent an article entitled "Famed NASA Astronaut Confirms Extraterrestrials Are Here." Have you read about Dr. Story Musgrave?*

Ah, yes, he is an incredible guy, incidentally. I think he has got three PhDs, at least two, and he has been in space for numerous hours as an astronaut . . . and he is convinced there is other life out there.

■ *He talked about seeing a craft that was approximately fifty to one hundred and fifty feet in diameter and he said (and I quote):*

> *Whether it was awash with debris or ice particles, I don't know, but it is characteristic of thousands of things that I have seen. What is not so characteristic is that it appears to come from nowhere and you would think that it is facing the dark side or facing a side that is toward you, which is not reflecting the sun. You would not think that you would see something there; it is really impressive.*

The article says that he is trying to communicate telepathically with ETs. I think it is just amazing, considering that he is a retired NASA astronaut. Very courageous of him to speak his truth!

Yes, he made several trips to space—an outstanding individual: one of the sharpest guys around. I am in awe of people like him, because he not only combined the book learning, so to speak, but the pilot capabilities and all the skills. We need to get more of these people to come forward, and we need the *New York Times,* the *Washington Post, 60 Minutes* . . . to bring their stories out.

Somebody who was connected with the major media told me, years ago, that they wouldn't touch UFOs with a ten-foot pole! One of the chapters in my new book is about public opinion and UFOs, and there has been a consistent picture, not only that there are more believers than nonbelievers. I hate that word, but I am stuck with it!

■ *I have read a number of polls, which indicate that people who do believe in extraterrestrial life and UFOs are consistently shown to be at about eighty percent of the population.*

Yes, it is a big number, even though the noisy negativists make it sound like only kooks, quacks, nuts, and little old ladies in tennis shoes believe in UFOs. I put myself out there. I have given over seven hundred lectures, in all fifty states, in nine Canadian provinces, and sixteen other countries and, like I said, I come on very strong.

The title of my talks is: "Flying Saucers ARE Real!" I start off by saying that some UFOs are alien spacecraft. I do that in all kinds of professional groups, engineering societies, and that includes universities, and I have had only eleven hecklers . . . and two of those were drunk! You are going to get more than that if you talk about sports, religion, or politics.

I commit myself—some people may say I should **be** committed, but anyway. . . . I spoke to a convention of Canadian journalists, held about sixty miles from my house in Saint John, New Brunswick. They asked me to give a lecture and I gave my usual hard-hitting, positive, strong lecture. I got a letter back from the guy who arranged it, saying, "I have never seen so many journalists change their mind about something so quickly, as when they saw how much data you had that they hadn't been aware of."

That is the kicker, I think.

■ *Let's remember that if we do have aliens arriving, it is going to dramatically change the way Jews see Arabs, blacks see whites, so many different groups that are at war (I've lost count!) . . . but, when it is earthlings and aliens interacting, it is going to be a whole new paradigm on the planet.*

For sure. Think how that affects your notion of religion, politics, economics, and art. These are things to think about. That's the idea.

■ *I think it is significant that this is coming so close to the time of 2012, which we all see as some kind of heralding of a new way for our civilization. We don't know what it is, but we certainly seem to pinpoint to this time on the calendar.*

I am an old guy, and I hope I am still around at that time to see what happens!

■ *It's just around the corner, now. And you will be around, Stan, because you have to shake people up; that's your job.*

That's my job . . . okay, I accept; I will continue to shake people up.

■ *And it's mine too. Helping people think outside the box of convention— beyond the matrix.*

■ ■ ■ ■ ■

Nuclear physicist Stanton T. Friedman holds BS and MS degrees in physics from the University of Chicago. He has distilled more than forty years of research on UFOs, sharing his work on a wide variety of classified advanced nuclear and space systems to his vast global audience. He authored *Top Secret/MAJIC* about Operation Majestic-12, and co-authored *Crash at Corona: The Definitive Study of the Roswell Incident.* He was the original civilian investigator of that very important event. He also co-authored *Captured! The Betty and Barney Hill UFO Experience.* His latest book, *Flying Saucers and Science,* was released in 2008.

Roswell:
The Truth Behind the "Myth"

■ ■ ■ ■ ■

Thomas J. Carey

Tom Carey is a man with a mission. After a stint in the Air Force, where he possessed a Top Secret/Crypto clearance, Tom became interested in anthropology and human evolution and received a master's degree in anthropology from California State University, Sacramento. He became interested in UFOs while in high school and rekindled that interest when he became the MUFON (Mutual UFO Network) state section director for southeastern Pennsylvania (1986–2002). Since 1991, Tom's research has focused solely on the "Roswell Incident" and the alleged retrieval/cover-up by the U.S. government of an alien spaceship and crew that crashed near the town of Roswell. This is the story of what really happened . . . as detailed in his book *Witness to Roswell,* co-authored with fellow investigator Donald Schmitt, in which they both have pieced together the Roswell reality, from thousands of interviews with people, dead and alive, who have kept the truth of Roswell alive. . . .

■ ■ ■ ■ ■

■ *I thought I had read a lot about Roswell, but I guess I hadn't read enough! There is so much information here and I have been involved nonstop since I read the first page; it really is a page-turner. The only problem I had with it was trying to keep track of all the key witnesses and data you bring forward—it is loaded with details and information. Let's begin our conversation with you giving us a little background as to what brought you to this mission of investigating Roswell, which eventually led you to write the book with Donald Schmitt.*

I was always interested in UFOs, as a lot of us are, from when I was a kid. I read a lot of Donald E. Keyhoe books and also books by Frank Edwards. When *The Roswell Incident,* by William Moore and Charles Berlitz, came out in 1980 (the first to be published on the subject), I got hooked on the idea of "crashed saucers." Crashed saucers were verboten: they were just not talked about in the UFO community, because in the 1950s there was a group that went by the name "Contactees." They claimed they had gone to Mars, Jupiter, and Venus and wherever, and these wild stories became associated with the idea of "crashed saucers" from outer space ... and weren't given much credence.

As the years went by, a UFO investigator in Cincinnati, by the name of Len Stringfield, accumulated a lot of stories from people who claimed they had either recovered crashed saucers, had seen them, or had been secondhand witnesses to them. The Roswell Incident occurred in 1947; by 1978, some of those people involved in that were starting to see the end of the tunnel—and started talking! One of them talked to Stanton Friedman, one of the original investigators on the case. That was in 1978: two years later, the first book—*The Roswell Incident*—came out, and I was hooked on that particular story. A few years went by and I finally got out of school—I was a subscriber to the *International UFO Reporter,* still in publication in Chicago, by the J. Allen Hynek Center.

I read about these two fellows, Kevin Randall and Don Schmitt, who were reopening the Roswell investigation. This was around 1990—about ten years after the first book came out. They talked about a group of archeologists from the University of Pennsylvania as being the ones who actually discovered the downed craft.

I thought to myself: "I'm here outside of Philadelphia, where the University of Pennsylvania is located, and I also have a background in archeology..." so, I called Randall and asked what he had done about these archeologists who had allegedly "found the ship." He told me that they had done some preliminary investigation, but that they "couldn't take it any further...."

So I said: "Let me have a crack at it," and he told me to "go to it."

I went to the university, looking to see what I could find out about these archeologists, and that is how I got started on this case, as an active investigator back in 1991—eighteen years ago. I can't believe it is now eighteen years later, and I am still on the case, certainly beyond the search for the archeologists.

■ *I would say so!*

Let's briefly outline the most salient details of the case—what exactly is assumed to have happened—that can bring us up to a basic level of aware-ness, before we get into the more specific details of your investigations.

The salient details are that on July 8, 1947, the Air Force issued a press release, stating that they had recovered a crashed flying saucer near the town of Roswell, New Mexico. They had been alerted to it by the rancher, "Mac" Brazel, who found it on his ranch, about seventy-five miles northwest of Roswell. The Air Force later came out with another press release that it wasn't a crashed flying saucer, but rather a downed weather balloon: this would be a rubber and tinfoil affair—your standard weather balloon.

This was a one-day story, around the world, that had its legs cut out from under it by that second press release, that stated it was a "weather balloon." Thirty years went by before the first person involved in that finally talked: Jesse Marcel, the base intelligence officer. He told Stanton Friedman that he had "held pieces of a flying saucer."

■ *I had Jesse Marcel Jr. on my show just a few weeks ago and he told me how his father had awakened him from a sound sleep in the middle of the night and dumped these pieces of metal all over the living room floor. He recounted having seen strange hieroglyphics on the metal.*

Yes, Major Marcel, according to Jesse Jr., was on his way back to the base with these artifacts, but he stopped home first, because he felt what he found was so exotic—"a spaceship from another planet," according to his son. That is why he stopped home: to show his family. By going public with Stanton Friedman in 1978, Jesse Marcel kick-started the civilian investigation of the Roswell Incident. We are now many years beyond that, and civilians are still investigating the Roswell Incident.

These are basically the salient points, without going into the numerous testimonies of our various witnesses.

■ *Yes, that would be difficult given the volume of testimonies you have, but we do just want to quickly describe, basically, how the story goes. Apparently, the rancher saw a strange flash and heard an explosion— and when he went the next day to investigate, he found that his sheep would not cross over a huge swath of the field, where he discovered the debris, to get to their drinking well.*

Yes, and we have witnesses all along the line: from hearing the muffled explosion (as Mac Brazel reported to his family the night that he heard it explode); to witnesses at the debris field; at the crash site; at the Roswell base; in the Roswell hospital; on the flights to Wright Field and the Fort Worth Army Air Field. We have witnesses all along the line, so we know pretty much what happened in this event. As you say, it was the rancher, the foreman of the J. B. Foster Ranch, who heard the explosion, and the next day he went out to check the sheep. In doing that, he came upon this large field, covered with this strange metal, which was scattered over about maybe three football fields in length and breadth.

He went into Roswell a few days later to report it. Mostly, he was interested in getting it cleaned up, because, as you stated, the sheep wouldn't cross to go to the watering hole. That is how the story got into the domain of the Air Force and also out to the public, because many civilians saw the crash site and saw the bodies; many were detained, and threatened with their lives. That is one prevalent thing that our investigation has uncovered: the extent to which the Air Force went, in order to silence civilians.

■ *In fact, I was going to ask you the logical question: why do you feel that it is okay to talk about this now, in such depth, revealing names of people and children of people, some of whom took their stories to their deathbeds out of fear of persecution and fear for their very lives? Why do you feel free to talk about it now?*

Well, why not? Many of them couldn't talk years ago because they did feel constrained by the menace and threats to their families' welfare—especially military personnel, who felt they were under some sort of security oath. Now, with the passage of time, many of the principals or the participants have passed away, and their children no longer feel such a threat to their families.

■ *And yet disclosure is still not on the table.*

Certainly not in America, but other countries are more open. America is not going to put it on the table, because it would be admitting to a lie. We know the government lies, but this would be a bold-faced lie that they would have to be admitting to now… a lie that has lasted sixty-two years.

■ *At least!*

Yes, at least sixty-two years. They are not prepared to do that, because there would be too many people with a price to pay, including people who were not directly involved in it, but certainly those involved in keeping the story quiet: politicians, especially. Politicians are not willing to pay the price. I don't look to them, shall we say, for strength of character.

■ *That is a sad understatement, if ever I heard one!*
One of the points I discovered in the book, that I did not know before I read it there, is that the squadron that dropped the atomic bomb on Nagasaki and Hiroshima was actually based at Roswell. I found that surprising and very, shall we say, "curious."

The 509th Bomb Group is the same group that dropped the two atomic bombs on Japan, in order to end World War II. The 509th Group was created strictly to drop the bombs in 1945, and trained exclusively for that mission. After the war, Roswell became our country's first SAC (Strategic Air Command) base. It was created to deliver the atomic bomb in time of war, so they relocated from Tinian Island, in the Pacific, to Roswell. The group was our elite military group in 1947—the most skilled officers and men, charged with the most important mission. To believe they would mistake a rubber weather balloon and a tinfoil radar target (held together with balsa wood and string) for an extraterrestrial craft is simply ludicrous.

■ *Yes, it really pushes one's imagination to the limit. It's also a very interesting "coincidence." We hear so many reports of craft hovering around nuclear facilities and military bases that I find it particularly interesting that a craft of ET origin would have been hovering around Roswell, home to the bomb squad, before crashing to Earth.*

New Mexico, in 1947, was the hub of our country's atomic activity. They had Los Alamos, up above Albuquerque, where the atomic bomb was basically created ... all the scientists up there; there was the Trinity Site at Alamogordo, where the first atomic bomb was detonated in 1945; they had an atomic weapons base, where they were developing atomic weapons, outside of Albuquerque. There was the White Sands Rocket Range, where the captured German B.IIs were tested.

So, a lot was going on in New Mexico. We don't like to speculate and we try to keep speculation to a minimum, but if I were coming in from the Andromeda Galaxy, wanting to check out Planet Earth, one of the places I would go to immediately would be New Mexico, because that was where all the action was taking place.

Roswell, being the home for the 509th Bomb Group, conducted regular tests for bomb runs. I should also mention that, in 1946, there was a combined Army-Navy operation called "Operation Crossroads." What it was designed to do was to test the effect of atomic explosion on surface ships. So, they gathered all these surface ships— surplus ships from World War II (including Japanese and German ships, and ships the Americans didn't want), on the Atoll of Bikini, way out on the Marshall Islands. Then, they dropped one atomic bomb, and a second bomb was self-detonated beneath the surface. The first one actually missed the target! One of the fins fell off and it missed the target, so the second one was not "dropped"—it was detonated, subsurface, and it literally obliterated the atoll.

Those photographs can still be seen on the History Channel. You can see bombs going off, and you see the big wave of water subsuming the surface ships.

■ *So many people forget that America is the only country that has ever used the atom bomb as a weapon.*

That's true! Certainly other countries have tested them....

■ *Yes, but the United States actually dropped this bomb on another nation.*

Right. We bombed Japan to end World War II, and the debate still goes on as to whether it should have been dropped or not.

■ *Hmmm.... I just can't find one good reason for dropping an atom bomb and killing hundreds of thousands of innocent people!*
I think that a lot of people are confused, like I was, as to the finding of the bodies from this craft. You point out very clearly that there were three sites. Before I read your book, I was under the impression that the craft skidded across the field and that they found it at the end of this long trajectory of debris. In fact, in the book, you describe how the rancher alluded to "the other site," where you believe he found dead bodies.

I had my daughter, who is a graphic artist, construct a map of New Mexico, detailing the three sites. As you say, it is "confusing." Don and I understand it, because we've been on the case twenty-two and eighteen years, respectively. So it is clear to us.

You have three sites—all in a line. The first site is where the ship actually exploded. We believe it was struck by lightning on the evening of July 3, 1947 (that's the muffled explosion that Mac and a number of his neighbors heard that evening and all the stuff rained down). That is above the field of strange metal that he found, which we call the "Debris Field Site," comprised of strictly wreckage: physical wreckage.

Two and a half miles east-southeast of that field, Mac found something **else.** We know this from the radio conversation he had with the radio announcer, Frank Joyce, the day he went into Roswell. We also know it from Dee Proctor, the little boy who was with him, the day he found the wreckage, son of his closest neighbors, Floyd and Loretta Proctor. It was there, years later, in 1994, that Dee took his sick mother, Loretta, to the second site. He thought she was going to die, because she had a blood clot in her neck, so he packed her up in a station wagon and out they went to the desert. There she was on her deathbed, he thought, and still he took her out on this bumpy ride in the desert to show her this other location.

He told her: "Here is where Mac found something **else.**"

Let's put two and two together: Mac Brazel tells Frank Joyce he found "little bodies somewhere else," and Dee Proctor tells his mother, "This is where Mac found something else." We have other witnesses who talked about actually seeing those "little bodies" on this small bluff, two and a half miles east-southeast of the Debris Field. We call that today the "Dee Proctor Body Site." What we believe is that, when the ship exploded, out flew three bodies. They were in the ship, unsecured, and when the ship exploded and disintegrated, out came these bodies and they landed at this site. That is the second of the three sites.

We have testimony from civilians and military alike of actually seeing an intact craft (what they saw was the egg-shaped inner cabin or an escape pod that was able to withstand the initial explosion) that stayed aloft or skidded along the ground at the Debris Field Site.

It most likely became airborne again, until it landed, about another twenty-five to thirty miles east-southeast of the Debris Field. There were bodies at that site, which we refer to as the "Impact Site."

■ *So there were actually six bodies found?*

We believe there was a minimum of four ... maximum seven.

■ *I see—that is another aspect that is confusing. Most stories describe finding three dead bodies and one still alive.*

The Impact Site is where the remainder of the ship came to Earth. We believe there were four there, as you say: three dead and one still alive. At the Dee Proctor Site, we believe there were between two to three bodies: all dead.

■ *In the book, I found it strange that the military called the mortuary looking for "several caskets for small children." Why would the military be asking civilians for children's caskets, when they were trying to cover this up, and why, I wondered, did they need so many—but now I understand there were more bodies than I originally thought.*

Who found the Impact Site? How does the story report this discovery, twenty-five miles away?

... by talking to witnesses. You know, Patricia, no one witness has the whole story. Each one knows a snippet and, for us, it's like putting together a jigsaw puzzle—putting the pieces into a framework that makes the most sense. We have witnesses to the Impact Site; as we said, we believe this was an inner cabin because it withstood the blast. By witness testimony, we know it was egg-shaped.

■ *How did it get into the hands of the military?*

It was discovered by civilians, who reported it to the Sheriff's Office in Roswell. Every call that goes into the Sheriff's Office automatically goes into the fire department, as well. The Sheriff's Office then called the military, because they thought it was an airplane crash and anything in the air always got reported to the base, as it was, after all, an air base. The sheriff reported it to the base and later on, he told his family that it was the "biggest mistake he had ever made in his life." He said that if he had it to do over again, he wouldn't have. But he did.

Also, the rancher came in to the Sheriff's Office—the sheriff interviewed him and when he called the base, he said, "I have someone here who claims he has parts of a flying saucer."

That is what brought Jesse Marcel into the Sheriff's Office. He sees the two boxes of strange metal that Mac Brazel has brought into the sheriff, and handles it, and knows immediately that this stuff isn't anything that we have on Earth. He takes it back to the base to Colonel Blanchard, who then dispatches Marcel and the counterintelligence officer, Sheridan Cavitt, to follow Brazel back to the ranch to see what's out there. That's how the military gets involved.

Now, that's all happening at the Debris Field Site. Brazel is out there with Marcel and Cavitt: they still don't know about the Impact Site. What happens is the group of archeologists out there doing their work come across this crashed vehicle—so they report it to the sheriff and the fire department. So again, the sheriff, Wilcox, reports this to the base. In the meantime, Wilcox goes out there, the fire department goes out, and while they're out there looking at it, the military arrives.

■ *And while they are looking at this pod, they see bodies?*

Absolutely.

■ *Phew! What a tense moment to be witnessing such a monumental scene—only to be confronted by the military.*

Yes, and they were threatened with their lives and told to remain silent.

■ *So the military go after the rancher. They take him to the base—what happens?*

They keep him for a week, at the so-called "guest house." Brazel didn't feel very much like a "guest." They kept him there and basically tried to get him to change his story, which he did, under intimidation and threat of reprisal to his family—perhaps with the aid of a bribe. There were reports that he bought a brand new pickup truck, shortly thereafter. They get Brazel (who had already reported the story that he saw the debris of a spaceship and little beings to Frank Joyce at the radio station) to retract his story. They march him around to the two radio stations and the two newspapers to tell a new story: that he made a "mistake." It wasn't a spaceship—it was a weather balloon. They take him back to the base and finally, after a week, they take him back to his ranch.

■ *. . . and this is where he has his new pickup truck and basically drives off into the sunset, right?*

Exactly. Shortly thereafter, he quit the ranch and went back to his home in Tularosa, New Mexico, and opened his own business. Here's a guy who never had two nickels to rub together, who has had it with being a foreman and being threatened, and he opens his own business in Tularosa. Something went on there that he benefited from financially. It took years for him to tell his family members, in dribbles and drabs, what happened. They all reported that he was very bitter, because he thought he was doing his "civic duty" in reporting what he had found. In the newspaper article that came out the next day, where he said it was a "weather balloon," at the end of the article he said: "Oh, by the way. I'm familiar with weather balloons and it wasn't a weather balloon ... and if anybody thinks I'm ever going to report anything to the military, ever again, they're nuts. It would have to be an atomic bomb before I would ever report anything again." He basically took back the story he had just told— he basically told the weather balloon story and then said, "Don't believe it."

He died in 1963, long before any of us ever heard about Roswell.

■ *What's so frustrating about this story—I know it must be for you as well—is that you just have to believe that a piece of this metal is somewhere hidden—all of these witnesses, two miles of debris ... surely somebody's got a piece of this. Of course, now our government has started to develop a form of this memory metal, as it has been identified, so even if it was found ...*

It's interesting that, of the civilians, the least cooperative have been the Corona ranchers. We think the fear of God was put into them by the military, because they went out there looking for these pieces you mention. In normal crashes, it was always the civilians who got there first, you know, looking for artifacts ... in this case, civilians got to both sites first. We talked to a number of witnesses who saw the wreckage, who actually handled the wreckage, and we know that, at one time, some of them had some pieces, but they all denied having them. The military came out and ransacked ranches looking for pieces of the metallic objects—they lifted up wooden floorboards, cut out the meal sacks used to feed the cattle, and even drained their water cisterns, thinking they might have hid the pieces in the bottom of these water tanks. They just did everything to these poor people.

Understandably, when people came calling decades later, they did not want to talk.

■ *Don't you just feel that there's a little piece or twenty, buried under the soil somewhere?*

Certainly under the soil—we certainly believe that the military didn't get it all, and that there still are pieces in the possession of civilians. We know that for sure. But we have not been able to secure a piece.

■ *How do you know that "for sure"?*

Because people we trust have told us that "they know somebody" who has a piece and they're just waiting for someone to pass away before they give it up....

■ *On the other hand, let's say the military has access to all these metal bits and have started to develop it. As I was starting to say before, we are hearing about memory metal now in our technology. So, in the event that we do see a piece, that it comes out, it may not be identifiable as "alien" substance.*

The memory metal is our holy grail. Don and I are looking for a piece of this original metal. This is the metal that is light as a feather, thin as the tinfoil in a cigarette pack. Those who handled it report that you couldn't cut it, scratch it, burn it, or deform it in any way. Yet, you could wad it up in your hand, to where it felt like you had nothing in your hand, hold it over a flat surface, say about a foot, let it go, and before it hit the surface, it would flatten right out to a flat piece, without a crease ... so that's our holy grail: memory metal.

What you have alluded to is something that we have just learned within the last six months, from an associate in Florida, Anthony

Bragalia, who has been investigating the Roswell case on his own. He started investigating what might have happened to the wreckage that was taken to the Wright-Patterson Air Force Base in Dayton, Ohio, which is the home of the foreign technology division, which tries to "reengineer" everything. If they capture a MiG-29 aircraft, for example, they try to back-engineer it. Tony was able to locate scientific papers, after 1948, which dealt with a contract that the Air Force had given to Battelle Memorial Institute in Columbus, Ohio, to try to reengineer this memory metal. The Air Force was able to determine, in their own labs, that it was made up of titanium and nickel—those were the two major elements of its makeup.

They were both pure grades of titanium and nickel, which we didn't have at that time, so they farmed it out to Battelle, which runs a number of labs in the country, and they came up with this nickel-titanium alloy that they call "nitinol," which is, in effect, memory metal. It's not as good as the memory metal retrieved from the craft, but it is still the same idea. Tony was able to trace these scientific articles back to right after the Roswell crash. There was nothing prior to the crash regarding experiments on nickel and titanium, or on what they called "morphing metal," or "shape recovery metal." Nothing! It's only after the Roswell crash, under these Air Force contracts. You put two and two together and it's pretty compelling.

■ *It is indeed compelling! We need to embrace the idea of ET presence, at the very least, so that we can start getting some of this ET technology that hopefully can help humankind and the planet—this is just an example of it. Surely a lot of our modern technology is back-engineered from things they've gotten their hands on.*

We're already using nitinol in medical devices. It's now being tested for the Space Program, because it has this morphing quality. Suppose you have a spaceship launch—or a telescope or something you put into space—and it gets hit by something out there. You don't want the skin of your device to get dented—but it can recover its form—you know what I mean? It's being tested for the Space Program. Something as mundane as your eyeglasses—the frames—if you notice many today can be twisted but return to the original shape—those are mostly made of titanium.

The genesis came from the Roswell crash.

■ *It is such an important development in the whole UFO/ET presence story. Let's get off the scientific track a moment here and get way out there. We can ask ourselves: are we getting all of our advanced technology from our own future selves, perhaps in another dimension? Are we getting it from ETs? There are so many hypotheses out there, all fascinating to contemplate. But here we have fact! This memory metal is so important to the story and like you said, nothing before Roswell and then, bingo!—all these people who reported seeing or touching this metal . . . and then, bingo again, suddenly we have laboratories developing metals that report to have these properties and now we have a form of it in our eyeglasses!*

And here's something interesting for your listeners, Patricia. It's not in the book, but we've learned of it since the book. I got an e-mail right after the book came out, from a fellow who was in one of the Battelle labs working on this stuff. He said they brought this stuff in, they said it was from Wright-Patterson Air Force Base, and they dumped it on the table and told him to figure out what it was. He said it looked like a bunch of junk.

He started describing the memory metal to me: how he could wad it up and it would do all these recovery things. The job was to figure out what it was made of, what it was, and how to engineer it.

We have a real live person who worked on that, as I said, since the book came out!

■ *That's very exciting! When did you acquire that little piece of information?*

About two weeks after our book came out, we got the e-mail. So we're going off in this direction with high hopes. He actually also sent me a photo of what he says he kept: a piece of the original wreckage.

I believe it is a photo of the original Roswell wreckage.

■ *This is an amazing revelation! Thank you for breaking this news here, with my audience. Are you going to be able to go look at this piece?*

Well, I have the photo! We have yet to decide what the venue will be to release it. I know I am going to show it in my presentation next week in Roswell.

■ *I wish I could be there—it will be an exciting moment!*

This is the first genuine piece of wreckage that I've ever seen.

■ *Again the question: do you think you will be able to actually touch this, in person?*

We haven't gotten that far yet. I'm just happy to have the photo.

■ *I would imagine so! No need to jump the gun, here!*

He kept it all these years and I asked him why; he said it was just something extraordinary. So it's a start. We thought, because we had a lead in this direction, that we would have a number of pieces in our possession by now. But often what happens is that people say they have something: a photograph, or a piece of physical evidence. One of two things always happens: either they get cold feet, or they are bogus from the onset, and in both cases, they just disappear on you.

■ *How do you know that this is not bogus? What is it about the picture that strikes you as unusual?*

For starters, his description of the work he did tracks exactly with what Tony Bragalia said **they** did. He uses scientific terms and processes that are way over my head, but what he described tracked pretty near to what Tony said. The lab he worked in is a real lab. So far, I have no reason to disbelieve him.

■ *This is kind of breathtaking for me—I certainly didn't expect you to announce this in our "daring conversation." What a gift!*

I have one more for you. This has also happened since the book came out: I'm getting chills just telling you about it. We have found the driver of the Air Force ambulance that drove the dead bodies from the Impact Site north of town—the third site. Since the book has come out, we have been in touch with the family of the driver, who drove those bodies from the Impact Site to the base hospital. It's an incredible story. The fellow (we have a yearbook from 1947, the only year they published a yearbook on the Roswell base of the airmen and -women) was in the yearbook—so we know we have the right guy. According to his family, he passed away, but he told them the whole story. We've been in touch with them, and the story is incredible.

■ *Tell us the story!!*

Well, he was called out to the site, where he saw three dead aliens, one alive, but he was told to put the dead ones in body bags. He described them. He drove them back to the base. He described the features pretty near to what everybody else who has seen them reported: they were three and a half feet tall; had very large heads for the frail frame; no nose . . . just two holes in front, where the nose would be; two holes on the sides of the head, where ears would be; the head was large like an inverted egg, sitting on a frail frame. They had four fingers, but the tips were like little smooth disks. He went into a lot of different things. He knew someone—one of the men who was with him at the time—who later committed suicide. It's a story that is still developing, but it is an incredible discovery for us.

■ *I'll tell you what, Tom. If I had a mystery that needed to be unraveled, I would certainly want you and Don on the case!*

We have over six hundred witnesses that we've interviewed (well, over a thousand); we have over six hundred that have attested that it was not a weather balloon, but rather an ET spaceship. From where it was, we don't know—but it crashed in 1947. The Air Force has zero. I repeat . . . the Air Force has zero witnesses to a "balloon event" in Mac Brazel's pasture—and they're sticking to the story of a balloon. But we're still finding new, key witnesses. Some may still not be alive, but their families are.

We are still on this case, Patricia. We're building the case, we're solving the case, one witness at a time, until we get into our hands an incontrovertible piece of memory metal.

■ *And it sounds like you're getting pretty close there, Tom.*

We think we are!

■ *But the thing is, of course, that the government can easily dismiss it, by saying that we have such metal in our laboratories. . . .*

No, the nitinol is trending in that direction, but it is certainly not as good as the original. The idea is the same; the idea came from Roswell.

■ *Okay, I am going to ask you a philosophical question, Tom, because I like to take these discussions to another level. Obviously, you're like a dog with a bone on this case, and you're not going to let go until you've got it. It's your life quest and I just love that about anyone who is so dedicated and who has such a clear mission. Question: what if you really crack this and you can prove, unquestionably, that this is a case of ET interaction with Earth?*

Well, Patricia, ever since we became human beings, however long ago, we've looked up at the sky and wondered: are we alone? Cracking the Roswell case will answer that question. We believe that we are not alone, and that someone has actually come here. There are many scientists, most I would say, all agree that there is other life in the Universe; the numbers are just statistically unarguable that other life exists. The argument comes as to whether any of that other life has actually visited Earth.

I believe it has, based on the Roswell case—there are also other indications, but my thing is the Roswell case. Knowing that, I don't think the financial institutions will collapse, any more than they already have. . . .

■ *Maybe they should collapse, so that they can reinvent themselves.*

What do you think is going to happen? Let's face it, something needs to happen quickly on Planet Earth. What do you think (and boy, I really do wish it for you—to be the one who announces it), when we have unequivocal truth of contact? What do you think will happen on this planet when the moment of proof is upon us?

I believe it will be arrival, en masse, of craft from other worlds. What do you think when, however it is announced to the human race, we have proof of alien presence on Earth?

It's a good question. Certainly, there will be days, if not weeks, of excitement—that's a given. I think there could very well be a medical and technological benefit, if the proof is in the form of a live being. But suppose the proof is simply the Air Force admitting that, yes, something crashed at Roswell and we know that there are visitors to our planet. If it is just admitting it like that, I don't think much will change, beyond what we already see: there will be excitement; there will be things to be gained, just from the knowledge of it.

■ *When President Clinton announced that we might have been looking at microbial life forms in the Martian rock sample, I thought that might have been the beginning of disclosure from the government.*

Yes, I know the rock you're talking about and they certainly looked like little critters to me!

■ *They did to me too!*

To me, it was: "There is life!"

■ *Yes, and it seemed that this was going to be the departure point from which they were going to now start releasing information, because no one was panicking over that rock. It was a great first step at disclosure, but it seems they decided to backtrack at the last minute. Why do you think they are not starting to release this information?*

... because to release it would be admitting to a lie that they have been perpetrating, ever since 1947. They would be admitting to having lied to us, and that's a big deal. No politician is going to cross that river and admit it.

■ *On the other hand, it seems every wrongdoing being perpetrated by our representatives gets dismissed under the "homeland security" umbrella. Anything: torture, misappropriation of funds, hijacking innocent people ... all for "security." Couldn't they just announce that it was withheld for our security and that, now that we are past this danger, they can bring it out? They're going to have to do it sometime—or, perhaps; the ETs will make their own announcement.*

They're also betting on there not being another crash! I don't see the government ever admitting this, unless dragged, kicking and screaming, to the truth.

■ *The truth is, something has to happen. With all the sightings going on around the planet now, it feels like we are absolutely at a point of reckoning with this most important issue, where we can once and for all recognize we are not orphans of the Universe.*

One thing I can guarantee is that there won't be rioting in the streets, like in the 1938 *War of the Worlds* radio show. We're sophisticated enough today, with many motion pictures depicting ETs, and our own space program, so that won't happen. And I think everyone will welcome the news ... I really do.

■ *Yes, and the polls say that something like eighty percent of the global populations believe in ET, anyway! It's time for the government to stop acting as our self-appointed babysitters, so that we can get on with the incredible task at hand of joining with our greater family of conscious beings.*

Well, you see what they did . . . they decided (in 1947 and reaffirmed in 1953) to suppress this information, because they felt that it was a national security risk, because they didn't want the lines of communication of our military establishments tied up with UFO reports. So they decided on this, halfway, to suppress information and ridicule it, so that they kill the reporting of UFOs. . . .

■ *. . . so that they could keep focusing on war.*

Well, that's what the military does.

My personal belief is that we are sophisticated enough today that people aren't going to be running out in the streets in panic. I don't think religious institutions or financial institutions would be affected. I think that life would soon get back to normal.

■ *. . . whatever "normal" is! What will happen when that great hour is upon us? We may soon know! Wouldn't we love to see Tom Carey and Don find that piece of metal that provides that very link?*

What an honor, Tom, to have heard of your latest breakthroughs first-hand and to have been able to talk to you, candidly, so far beyond the matrix.

I am honored, Patricia, to have been a guest on your show. And I want you to rest assured:

We are on this case to the finish!

■ ■ ■ ■ ■

Read more about Tom Carey at:
www.roswellinvestigator.com

The book is:
Witness to Roswell: Unmasking the Government's Biggest Cover-up

Disclosure Now!

■ ■ ■ ■ ■

Stephen Bassett

The leading advocate for ending the sixty-two-year, government-imposed Truth Embargo regarding the extraterrestrial presence on Earth. Stephen Bassett is the executive director of the Paradigm Research Group, seeking and demanding answers to the unanswered questions—the truth of contact.

■ ■ ■ ■ ■

■ *Thank you, Stephen, for dedicating your time to talk with me to elucidate what you are doing to help convince the U.S. government to disclose the ET presence on Planet Earth. We're going to bring people up to date on all the hard work you are doing to blow the lid off this ... what shall we call it?—"Great Cover-up."*

We call it a "Truth Embargo." Truth Embargo refers to the formal effort/policy of the government, instituted in the late 1940s; I think it was fully in place by 1953. The reality of the phenomenon that it turned up, which we call now the "UFO phenomenon" (a very crude term to describe it), has been in play for a very long time. That reality had to be embargoed, because it could not become consensus reality. They did this for national security reasons. It's difficult to argue that it was illegal (it certainly wasn't illegal), so that is why I don't call it a "cover-up."

The time period was extraordinary. The Cold War was starting to shape up. There was huge concern within the government, both in the U.S. and with our allies, that the Soviet Union had gotten bomb secrets through their own spies.

We had withheld that program from them. That was a mistake on our part, perhaps; obviously, it didn't instill any trust in us from the Soviet Union. The ideological break was there. They had German scientists, like we did, who were passing on missile development secrets, and we were going to build missiles, we were going to build bombs. It did not look good at all.

In the midst of all that, there were extraterrestrials literally climbing out of the sky! They had known that something was afoot during World War II, because of the foo fighters; I think they already knew that there was an extraterrestrial presence at that time, but they couldn't do anything about it. In the middle of a world war, what are you going to do about UFOs if your planes are already engaged?

■ *Right! We certainly would want to interrupt a war for something as important as extraterrestrial contact!*

Well, right.... It probably would have been a good idea if they had, but they didn't. They made the decision: "Until we understand what is going on here, we keep quiet."

They did have technology by 1947 at least; they may have had it sooner, from a vehicle that crashed (maybe more than one) or was **shot down.** They had technology that needed to be understood, so they had to deal with this issue. I am sure that the intention, at some point, was to eventually let the people know, if they needed to know more. But for then, they embargoed it.

■ *Let's be sure we are clear who you are talking about when you say, "they."*

I'm referring to the U.S. government ... the military: the people inside our government that had to deal with our national security. It wasn't a secret, because these craft were flying all over the world. I mean, what are you going to do about that, right? What they had to do was to maintain control over it, so this was a far more complex and difficult task, but they ended up being successful.

That's what the Truth Embargo was about—embargoing even the acknowledgment of this issue as "real" to the people, with the intention, undoubtedly, of revealing it sometime later. What happened then was that the Cold War just got worse and worse. There was the crisis in Berlin, which led to the Berlin airlift; there was, of course, the Cuban Missile Crisis in 1962; the Vietnam War got under way a few years later; we were fighting a war on the border of China ...

■ *We never stop having wars, let's make that assertion right off! Is there ever a time on this planet when there is no war?*

No, not that we can recall. Humans have been at war with themselves for almost since the Flood ... and eventually went to war with the planet **itself.**

That's what we do; that's how we roll.

The Cold War got worse and they felt, I am pretty sure, that bringing the ET issue out and acknowledging the extraterrestrial presence, while ten thousand nukes are pointed at each other across both polar ice caps, was probably too risky. It could have destabilized what was already a bad situation and led to some paranoiac pre-emptive strike. I think (and I'm just guessing, but it's a good guess!) that the decision was made that, until the Cold War was over, this could not come out. They had a very powerful motive to maintain the Truth Embargo, and they did that through disruption of research; disinformation; planting of false stories; intimidation of activists; misrepresentation . . . you name it! They did it.

They had unlimited funding to do that and, of course, they had the license to do it, as part of the defense of the United States against the Soviet Union. The embargo meant that the truth was not going to break until the Cold War ended, which did end, formally in 1981, on Christmas Day, when the Communist Party of the Soviet Union formally dismantled.

At that point, the potential for disclosure dramatically increased. But it was not a simple matter, meaning, "Okay, the Cold War's over—I guess it's time to tell everybody about the extraterrestrial situation!" They needed time to get their ducks in a row and their act together, and already we had post–Cold War "issues." And there was another important fact: the military intelligence community was very concerned about who the president was.

After all, this was to be the biggest event in human history. It had global implications unlike probably anything that has ever come down the pike, and so the president's capacities, and his relationship to military intelligence and the national security community, was obviously going to be fairly important.

■ *So it's a good thing we didn't have disclosure with George W. Bush!*

You're getting the idea here! His father, George H. W. Bush, would have been fine. He was like the quintessential insider, the national security president. He had all the credentials and he had all the connections. He would have been "perfect." I think they started to plan to do something in his second term. The Cold War ended in the middle of George Bush Sr.'s first term, and then the Kuwait War started right thereafter. It was just not possible to get this out in his first term. I do believe they were getting ready to get this done in his second term, which would have started in 1993. I think that's why things kind of loosened up around that time—I think Corso emerged shortly after that. He wrote the book *The Day After Roswell.*

Unfortunately, for the disclosure process, Bush Sr. did not win a second term. For reasons that are not fully understood (they think his health had gone bad, but this has never been publicly acknowledged), he simply wasn't up to another term.

■ *Yes, he did fall off that platform in Japan. That was pretty devastating for his presidency, as I recall, even though it was underplayed by the media. It really did raise some flags.*

That was an embarrassing moment, but it was attributed to food poisoning. Near the end of his campaign in 1992, he was definitely not on his game, and he lost to Bill Clinton. One of the great ironies of history is that Bill Clinton was exactly the opposite to Bush, in terms of his relationship with the national security structures.

They did not like him; they did not want him; they even despised him.

■ *How do you know this information?*

It was all in the news! You had members of the military openly castigating the president! Top military had to actually send out memos saying it was "treason" to openly castigate the president, while you are in uniform, in front of your troops. He was viewed as a pot-smoking womanizer, draft dodger; he was just an ephemeron to the military-intelligence complex, and they wanted nothing to do with him.

The first thing he did, when he got in, was to try to overturn the ban on gays in the military, and they came back at him very strongly on that.

He was going to get no slack at all; he was the last person they wanted. He was viewed as a "leftist" and a liar. Remember . . . the intelligence community develops complete dossiers on anybody that gets near the White House. A lot of people don't know this.

The only qualification that you must have to be president of the United States is to be thirty-five years of age and a national American citizen, born on U.S. soil—that's it! You don't have to pass a national security test and you don't have to have national security clearance, so they start collecting information on anyone who has intentions of becoming president, and of course, we're talking about the full might and force of the U.S. intelligence community.

These dossiers are never to be used unless national security is at threat, meaning that if they were to do a background check on a candidate in an election, and discover, for example, that this person had molested children, but had never been caught, it would then very possibly come out. But it would take a lot for this kind of information to be released. These dossiers are collected, as a matter of national security, but they are never supposed to be used. However, the people who collect these dossiers get a lot of information about a candidate, so there was a lot known about Bill Clinton when he came in.

I have a hunch that they broke protocol, and shared some of that information with the right wing, which went after the administration almost immediately. In any event, Clinton was not going to be the disclosure president. Every effort was made to remove him from office. It did not happen; he went on to a second term. So, actually, disclosure was off the table for the eight years of his presidency. Then, of course, we had a new election and a new president.

Some people may be shocked to hear that George W. Bush was not acceptable to the military-intelligence community either! He didn't have the background; he didn't have the skills, nor the intellect, in their view, to address this issue. He got into office, primarily, because of the liaison he had made with the Evangelical Christian Right, which started while he was working on his father's campaign. His father had no interest at all in the Christian Right. George W. did. He embraced them ... this is all written up. This is what helped him get into office. The only asset that he had, in terms of national security credentials, was his father.

In fact, he was at breach with his father almost immediately, and that breach got worse and worse as he moved toward the Iraq War. So, again, disclosure was off the table with George as president. The Democrats were convinced that they would win in 2004, and I believe the Democrats made a decision, around 2001, that when they got back into power, they would get this done.

What am I saying here? This is what people need to know: most of the people in government today, people in high positions, know there is an extraterrestrial presence. The fact that they don't tell us—that they don't issue press releases or talk about it before the microphone—doesn't change that. These are smart, intelligent people and they're "connected." The higher up you go, the more connections you have to the military-intelligence community.

■ *What I hear you saying here is that it is the military-intelligence community running things, not the president.*

Well, no—I'm not saying that. The military-intelligence community has a major role in the running of the U.S., and they do have impact on other nations, although that impact has diminished profoundly in the last ten years. A lot of people run the government: let's say it's a "group effort."

■ *Right—but it's not up to the United States alone to make this disclosure. We've got plenty of other nations in the world who have their own investigatory bodies and networks, who could also be coming out with this information. Isn't there a "higher authority" they are all beholden to?*

All the First World governments know there is an extraterrestrial presence. Does that mean that everybody in the government has been taken to hangars and shown bodies from these crashed vehicles? No. Have they been given extensive briefings? No. Are they aware, however? Yes. Have they "figured" it out on their own? Yes. Have they been "tipped off"? Of course!

You don't have to be shown bodies from alien craft to know that it's real. There is the really extraordinary disconnect between reality and perception. People think that this is a secret and everybody's in the dark. Not really . . . **they know.** The ET presence is pretty well known worldwide, but they haven't acknowledged it yet. We're in this sort of purgatory—a transition period—and this disconnect reflects how powerful this is. We've gotten used to this—we've seen aliens on TV and movies for decades. Most people, on some level, realize it is true . . . but, of course, it is still a big deal. This transition to the open acknowledgment of the ET reality is a big deal, and it's going to take some time.

It has taken too long already.

The citizens' movement to bring this to the public is picking up speed: the disclosure movement is gaining power. Witnesses are coming forward; books are being written; documentaries are being made; the Web has come on the scene. . . .

■ *And many more really credible people, like the astronaut Dr. Edgar Mitchell, and many military people—credentialed people—are openly discussing seeing craft in detail.*

Yes, and one of the reasons they are coming forward is that principal barrier to disclosure—that logical and perhaps appropriate basis for maintaining the embargo, based on the fact that the Soviet Union and the U.S. were always on the threshold of a nuclear war—was gone. Now people wanted to talk about this: they wanted to get it off their chests, as it were, and they felt they could now. Not everybody, though; many are still under some kind of national security oath or agreement, based upon their position, and they're not allowed to talk about it. But some are not, and a small percentage of them have started coming forward. These numbers have grown. So, the pressure on the embargo has grown substantially, particularly since 2001, but again, President Clinton was not acceptable as the disclosure president.

So, here is the situation we are in now. As I said, I think that the Democrats decided that, when they got back into power, they would do this. Well, they have just gotten back into power. Time now has passed, since the end of the Cold War; the pressure on the Truth Embargo is worldwide and enormous, coming from many directions. The embargo is basically being shredded. The government really stopped what I would call "formal" intervention; those formal efforts to disrupt, misinform, and intimidate stopped around 1999. I think they knew that something could happen at any moment, and another country might break ranks, and obviously they didn't want to be caught *"inelegant,"* deliberately screwing around with this issue—so it was best to step back.

In some respects, then, the embargo moves on from 1999 under its own momentum.

Now, it's 2009—the Democrats come to power and we have a new president, Barack Obama. What that means is that the window for disclosure has opened as wide as it has ever been, since 1947.

We are on the threshold, and **it could happen at any moment.** This is what your audience needs to understand.

■ *This is what I have been telling people and writing about, that certainly contact is imminent. We can feel the tension in the air, with regard to this phenomenon: it's a huge reality and it seems, especially now, to really be breaking. We have fleets of craft coming over the horizon in England and other locations. Sightings are everywhere around the planet, more and more often.*

So are you suggesting that Obama has been selected to be the disclosure president?

I think it would happen regardless, even if Hillary Clinton had won, even John McCain.

■ *Hold on! Let's backtrack just a moment here. What about John F. Kennedy? According to the story that prevails, he supposedly stated that he was going to "bring the truth to the American people on the question of extraterrestrial presence . . ." just a week before he was assassinated.*

Yes, there is some substance there that Kennedy was thinking along these lines.

■ *More than thinking about it—didn't he issue a memo to this regard?*

Yes, there is that memo—it is out there. You have to read between the lines in that memo, but let's just say that it wouldn't surprise me at all if Kennedy had been planning to make a move, and let me add that there was no way in hell that that was going to happen! The military-intelligence community was not going to allow the Truth Embargo to end in 1962.

■ *Do you think that might have had something to do with his assassination?*

It's possible—but it's pure conjecture. We just don't have that evidence. And we know there were a whole lot of people who had a whole lot of reasons to end his presidency and so, who knows?

... Back to the present. My colleagues and I have the job of scrutinizing all of this very closely. We're sort of like the people up in the crow's nest of a ship. Our job is to be up there watching for icebergs. The people on the ship are partying, or in the restaurants having dinner, maybe in the casinos or sunning on the deck—so they're not looking for icebergs. We're going to see them first.

We do get people wondering, "How do you know that?" That's because our job is to **pay attention**.

■ *Can you tell me more about this organization, so that people who are asking those questions will know a little bit more about who you are and who the people are in this organization? We know that you are a lobbyist, an activist, and a columnist—what about the others?*

There is a disclosure movement under way that involves a lot of different organizations, involving a lot of people. One of these organizations is Paradigm Research Group, which I founded in 1996. Its fundamental goal (there's a mission statement that explains it) is to advocate for the resolution of the UFO issue politically: the political resolution of the UFO situation, which is disclosure. Without disclosure, you have not even begun to resolve this issue. We've been at it for thirteen years, and along the way we've networked with a whole bunch of other groups; if you go to www.paradigmresearchgroup.org, you can find links to these other organizations, groups, and projects. There are scores of them.

I have helped to build a network and to network with other groups as a movement—but it's in cyberspace. It doesn't have a big office in downtown Washington. We don't get hundreds of thousands of people and bring them into the capital to march up and down the mall—that's difficult to do now. It would cost too much money, and this is not an issue that motivates people in that way. It is a movement that is built in cyberspace, where all political movements will be built in the future.

The disclosure movement is watching what is going on, and what I have to tell you, again, is that we are right on the threshold of finally disclosing this truth to the world. Of course, it's not a guarantee or a "fait accompli."

One of my contributions to this was to create a project called the "Million Fax on Washington." It was planned in early 2008, at the very beginning of the more formal campaigns, based first on the premise that whoever won the election would be considered acceptable. The further along the year went, the clearer that seemed: Obama, Clinton, or McCain would have been acceptable.

Secondly, this was going to be the longest and most expensive presidential campaign in history, involving scores of candidates, scores of debates, and huge amounts of political coverage.

And then, the ET issue was much more mature. The idea was to get that issue into play: get it into the coverage; hook it into the coverage of the presidential election; hook some of the candidates themselves to the issue, legitimately. The idea was to really get it out there, including questions asked of the candidates about the UFO/ET issue, which actually happened. It didn't happen until September, but it did finally happen.

We accomplished that.

The other thing was to continue to build the disclosure network, to make it stronger and as powerful as possible, and a lot of progress has been made there. All of this was a prelude to the "Million Fax on Washington," which was launched in October, to start on November 5, 2008.

The concept was very simple. There was going to be a new president. We needed everyone to send that new president a fax, e-mail, or letter calling for that president to resolve the issue: to support hearings; demand briefings; bring out the technology (to get that ET technology out of the government labs that they have been working on for sixty years). We think the physics of this technology could be significant in addressing the problems that humanity has today. In short, we needed people to demand disclosure. We wanted to get those letters in during the days of transition—the seventy-seven transitional days between the election and inauguration—and that has happened. Thousands of letters, e-mails, and faxes went into the transition headquarters.

Next, Phase Two was launched to then redirect that correspondence to the White House starting on January 21st for the first one hundred days of the administration. That was extended to one hundred and thirty-one days, because their plate was extraordinarily full the day they walked into the White House. So, the correspondence began flooding into the White House and that went on right up to May 31st.

The White House has yet to comment on this, but we expected that.

That kind of set the stage for the third phase, which started a couple of days ago (June 1, 2009): the critical phase. Everything going back to 2008 that PRG was involved in led up to June 1st, to what is happening right now (at the time of this interview).

Something else we did following the election: using the media as the conduit, we got a message to the White House that was pretty straightforward:

You need to end the Truth Embargo as soon as possible during your administration for three very important reasons:

1. If you don't do it, very soon your name is going to be added to it. This is going to be **your** embargo, as it was for the president before you, the president before him, and so on, going back to the times of Truman. When that happens, you have a problem—not the least of which is that, by not ending this embargo, you will completely put the lie to one of the major pillars of your reformist agenda, of an "open and transparent government." It will destroy it and wipe it out. So, you have a serious issue here. You need to do it.

2. We have been told every day, for quite a while now, that we are facing enormous global crises: economically, humanely, threatening wars over water and food, dying children, poverty and the terrorism that it generates, climate and environmental crises. The solutions are meager and yet all manner of dire consequences are being tossed at us. Some people call it "fearmongering"; some call it "reality-based thinking." Whatever… the point is we have to have every club in the bag, if we're going to par this course. ET technology cannot remain sequestered behind the Truth Embargo.

3. (This may be the one that gets their attention the most.) It's clear that other nations are breaking rank with the U.S. on the ET issue. They're not willing to go along with us, or follow our lead on this much longer. If our government does not disclose the ET presence soon, I think that there is better than a fifty-fifty chance that another nation will do it before the end of the year, and then the U.S. will basically be irrelevant.

■ *This is what I brought up earlier—the fact that the U.S. certainly doesn't have exclusivity on the information.*

Yes, our credibility and world esteem have fallen dramatically over the last ten years and it truthfully started falling after Vietnam—it's been eroding. Up to Vietnam, our prestige in the world was extraordinary. With the Marshall Plan successfully defeating fascism, staving off the communist threat, we were perceived as the defender of freedom in the world. From Vietnam forward, we've basically fallen off the cliff.

There was a time that when America said "jump," the rest of the world said, "How high?"

That ship has sailed.

■ *I agree that, unquestionably, the number three scenario of your list, all brilliant, really brings urgency to the disclosure question for the administration.*

To whomever it matters that the U.S. is first on this, it will be the greatest political legacy of all time—for whatever administration is behind disclosure—unless, of course, numerous countries get together and decide to toss this over to the UN.

■ *Stanton Friedman and I discussed disclosure at length and he wisely said: "The problem is we don't really have a leader who can speak for the entire human race: that would not be the president of the United States, nor would it be any of the governmental leaders of this world."*

From what I see, none of these leaders has the greatness to speak for the whole of humanity.

I don't know if anybody ever put it quite like that before . . . let's not think of one person.

■ *Okay, but let's not think of the UN either, because we know that is an ineffective organization. So who have we got?*

Yes, that's my point. The UN has been diminished and undermined, and it is not properly positioned to do this. Therefore, it is going to be a nation. A nation will start it and the question is: "Will it be the U.S. or not?"

Again, for whatever nation does this first, it will be highly significant.

■ *To say the least!*

There are some pragmatic reasons for why the United States will probably do this. First, we have lost enormous respect around the world and we have broken trust with the people. The only way you can get that trust back is by getting into the truth business. If the U.S. ends the Truth Embargo, it will be a huge step forward in normalizing the social contract in America, as well as our relations overseas.

In other words: "We have kept this from you, but now we're telling you...."

If another nation does this, we get all the downside. If we were the ones who withheld this for sixty years, and we're not the ones who bring it out, it would diminish the U.S. a lot and, of course, make it difficult to then take the lead on the issue. I think if the U.S. gets its act together, it could lead—and lead well.

Frankly, in the long run, this is not going to matter. The planet is going to learn about ET presence, one way or the other. Down the line, it won't matter that much which nation started it.

■ *Yes, let's de-politicize the question just a moment and talk about the impact on the human race. Whoever is the announcing party, most likely the announcement will have to have been generated by some big event, like what I think is coming (which of course can be considered conjecture): the possibility of massive sightings all over the globe—where there can be no more denial. Even now, daily, we are hearing of remarkable sightings—fifty craft here, sixty there . . . appearing overhead. Whether or not we've got a politician that appropriately makes disclosure, it looks like the human race is about ready to be confronted with this reality.*

What happens then, in your opinion, Stephen?

My position on that is fairly clear. We do not want a traumatic disclosure process: the more dramatic it is, the more destructive it will be. So, this idea that we need these mass sightings to drive the government is not good. We need the government to do this in a proactive way, without the ETs becoming overtly aggressive in terms of what they are doing up there.

Let's take a classic example. They park a mother ship over the capital, two miles across. I guess that's the ball game, isn't it?

That's the worst-case scenario, because, essentially, the government is then intimidated into disclosure. The violation of the social contract has now been made completely clear and broken. The perception of the people is that the State is powerless: the State has violated the social contract and is now being cowered into telling the truth, by a massive mother ship hanging overhead. This is incredibly traumatic to the relationships between all peoples and their governments worldwide. It would be very destructive, and it would create a very problematic post-disclosure period. So, we don't want that to happen.

■ *True, but on the other hand, there are other scenarios of arrival, one of which is not a menacing mother ship over the capital, but rather more and more sightings, which are indeed happening now, continuously appearing. This allows people to get used to the idea, and nobody is refuting it. The governments are leaking it more to the press; the governments are saying it is "possible." It appears that this is a joint effort. More sightings, more press about it—it seems like we are being prepared for this.*

That's exactly what it is. Now whether that's a "joint" effort, or whether it's incombinant, we don't know. But yes, that is what is going on—they are being seen, and it helps to drive the terrestrial process to disclosure.

■ *A lot of military people are coming out and describing their experiences and what they have seen and know. Why would they be allowed to do that if the government were intent upon prolonging the Truth Embargo? It looks like we are being prepared for that big day of disclosure.*

Yes, the disclosure process is on the way. I am just suggesting that the dramatic event needs to be avoided: a threshold event that is ET-driven. We need this process to be as terrestrial as possible. The ETs seem intent upon helping us along, by being less than discreet in their comings and goings. Putting a crop circle down in England is kind of a nice little nudge . . . so that's where it is and I hope it stays that way. But the ETs can lose patience, and they can do what they want. If they want to park mother ships over the capital, they can do that anytime they want to.

■ *I would assume that, if they wanted that kind of shock effect, they would have done it by now.*

Again, everyone's patience has a limit. I happen to believe that there is a reason why disclosure has to happen now. It has to happen pretty soon, and it's very possible the ETs feel the same way. If we continue to drag our feet and find excuses to put it off, they may be

forced to do something dramatic. There may very well be a time factor here. And I reiterate—that would be bad.

■ *In referring to the time factor, are you referring to the perilous condi-*
tion of the Earth?

Who knows?

■ *You just said you have an idea of why it has to happen now. . . .*

Yes, I have an idea but I can't specifically say what. We are up against
the wall in a lot of areas. I can't point at one of them and say, we
have to disclose by a certain time. I am simply suggesting that I
sense a time-delineating factor here, that at some time the ETs may
have to act.

Now, this is a very complex argument. Not many people are going
to get this—and that's okay. The point is this: the sense of urgency
is clear amongst the First World nations. I don't think anyone can
deny that. There is this powerful sense of urgency that has emerged,
on a collective basis, that is unprecedented in history, where you
have scores of nations and billions of people who are hopefully
appropriately (not from fearmongering and manipulation) in heightened states of urgency, regarding the status of the human race and
the planet itself.

This is really very profound. It is always profound when a large
group of people has a certain similar sense on something, because
humans are unbelievably diverse in their opinions. But, right now,
there are a whole lot of people . . . we have their attention. Their
attention is focused on the fact that we are running out of water,
we're running out of fish, out of oxygen . . . we're running out of ice
caps—running out of ozone protection. . . . We're running out of **time.**

And, of course, we still have twenty thousand nukes that could still be used!

There's a sense of urgency. Whether that sense of urgency is appropriately connected to this whole ET trajectory or not, I don't know—but my guess is that it is.

I suppose what will be very important for us all to consider is that we not glorify the ETs as our next "saviors," whose job it is to save us from ourselves. A lot of people seem to be getting stuck in that mindset. Some really believe that, any minute, the mother ships are going to swoop down, pluck out the "deserving" humans, and then take them away from the Earth, so they can go and destroy another planet, or maybe to another dimension ... even "heaven" itself.

Inevitably, it is clear, we could use assistance, because we are surely not doing a very good job at planet management. But let's be responsible for what we are creating on Planet Earth.

A good analogy is to think of the Earth as an adolescent—about a thirteen-year-old. The principal job of the parent is to get the kid from thirteen to the age of twenty or so without killing him- or herself, or ending up in prison, or addicted to drugs, or whatever awful things that young adults can get into, because they don't know better.

As long as we are a self-extinguishing society, the question remains as to whether ETs might want to intervene to help us, or leave us to our own evolutionary process. But at the moment we become dangerous to the greater galaxy, with behaviors that could dramatically affect life beyond our atmosphere, then this becomes a point of interest for intelligent beings that are outside of our realm.

I don't really think we have the ability to affect much outside of our planet, so I'm not really onboard with that idea. But we most definitely have the ability to screw up the planet! It wouldn't surprise me at all that sentient life (once it is pretty much advanced), wherever it exists, has a powerful appreciation of a biosphere.

When we see the planet from space, we are moved—that is a legitimate emotional reaction. This being, once a ball of gases and magma billions of years ago, is now this water-covered, atmosphere-enveloped blue dot: this incredible place where life thrives. We know that, while there may in fact be billions of such biospheres in the galaxy (as recently stated by scientists in the States and Great Britain), that this is still an extraordinary, special place, as any biosphere would be. I believe that sentient life would have a certain appreciation of that and so the destruction of that biosphere would be something that they would oppose. . . .

■ *Wouldn't you agree that the destruction of that biosphere would affect them dramatically, because everything is interconnected, throughout the galaxy and throughout the Universe? Aren't we all One? Anything that would destroy a planet would have to have repercussions beyond the immediate biosphere.*

That's a nice sentiment and it could well be true. I don't know. It is not my area of expertise.

■ *Your point is that ETs would be concerned about earth events strictly in the case of the destruction of the planet?*

Yes, they are completely concerned about earth events. The evidence for that is fairly overwhelming. What the motivation for that concern is, and where that concern is going, we don't fully know yet. Nevertheless, **they're here.** They're here at a rather critical time and I don't think that's an accident.

There's the setting for what is about to happen.

■ *Earlier you mentioned the crop circles. I would like to ask you how you feel about this phenomenon, with regard to what they represent. Clearly we have many different opinions about it.*

What they represent, I don't know. My view about the crop circles is really simple: when this thing first got under way, they were predominantly "extraterrestrial." Over time, the percentage of crop circles that are made by humans grew, so now there's a balance between the percentage that are human-made and the percentage that is of extraterrestrial origin. It's become very complicated, but by the same token, in a way, it's very symbolic (whether it is intentional or not) that there is almost a sort of dialogue going on between humans and extraterrestrials.

In the crop formations of England, the locations are no accident. The one message that I get from crop circles (and it's an extremely important message) is that this is the most dramatic and compelling evidence that we can look to for the ultimate outcome of this, which helps us realize it is not going to be bad ... that we will be relatively pleased with the post-disclosure reality, and even with formal contact. Whatever these beings' agendas are individually, in the collective they are not destructive.

I think crop circles are the best thing we've got to understand this ... in fact, that's what I turn to when I worry a little bit. There's evidence that there could be some unpleasantness coming; how bad that could be we don't really know. The crop circles are a good indication that it's not going to be bad.

It's a wonderful phenomenon, and I really kind of envy the people that are involved in it, because I think it is a much more satisfying and emotional experience for people. Let's put it this way: I'd much rather be a crop circle investigator than a cattle mutilation researcher, for obvious reasons.

▨ *I go every year to the crop circles in England, and, in fact, I will be leaving for the fields soon again this year. The formations of this year, 2009, are pretty incredible. I sent you a picture of a recent formation—the Jellyfish—which is over six hundred feet long.*

Yes, it is really impressive.

▨ *It is. And there's not a footprint in the wheat or broken stalks anywhere near it (or at least not at the time of the aerial photo)—it is just breathtaking!*

Were they able to confirm that this was done overnight?

▨ *I can find out for you. I thought this came down during the day, but I have still not researched it. I will get back to you with more information about it.*

Well, if that came down in less than twenty-four hours, then it is almost certainly extraterrestrial.

▨ *But I agree with you—if humans are doing these master works, then great—we most definitely have a dialogue going on.*

Here's what needs to happen now—we need to wrap on the Truth Embargo. We need to end it. I think the government plans to do it, but the government is also famous for getting clay feet and losing its nerve.

We have to ensure that this gets done, so the final phase of the Fax on Washington is designed to do that. This is where we are: we are going to bring the hammer down, in a way. In Phase Three, we shift the focus from the White House to the White House press corps, so when you go to www.faxonwashington.org, what you find is now all the information necessary to direct correspondence to the White House press corps, c/o the White House Correspondents' Association.

What do we want heading their way? Two things:

1. Everybody who sent a letter to the president about disclosure over the last two hundred days or so needs to copy that letter to the White House press corps, c/o the WHCA. They need to get a lot of these, so that they have tactile proof, in their hands, that the White House is getting this correspondence.

2. We need everybody, whether they sent a letter previously or not, to write a letter, fax, or e-mail now to the White House press corps, saying: "We want you to start asking about this issue now. We want real questions—we want real answers. You need to start immediately."

At the www.faxonwashington.org Web site, you will see a list of fifteen carefully selected questions that are hooked to the political realities of our time. They are also hooked to members of the administration and what we know about their connection to the issue. There are links to documents, research, video clips—the works! There is a ton of information there that backs up every one of those questions. We know the answers, in a way, but we want the administration and people in the administration to respond. They have never been asked these questions—that's why this thing keeps on going.

If even a handful of press were to ask the appropriate questions and require the appropriate answers, it would quickly escalate into a media firestorm and it would burn the Truth Embargo to the ground.

■ *Of course, most of the media are afraid to ask questions, until they are sure they will be allowed to ask those questions.*

So we're giving them permission! We need tens of thousands here—with letters going into their offices. In the last four days, I've already talked to about six or seven million people, so it's a question of whether they want to reveal this truth or not. If you don't really want to know if there is an ET presence ... if you don't want this acknowledged ... if you don't want that technology ... if you want to be paying six dollars a gallon for gas, and are content to sit there watching the world fall apart around you, then don't do anything. Keep watching your television.

If, instead, you want this truth out—if you want the government to get back into the truth business, if you want to know about ETs, if you want that technology available, so that maybe you'll be paying only a couple of dollars a month for your air-conditioning bill in a few years, then you need to write a letter to the press corps. This is the final phase.

■ *And let us not forget: if you want to live the magic of being part of a civilization that finally globally recognizes that we are not alone, do something about it to make this happen sooner, not later.*

Please help Stephen and the people who are supporting him by contacting him through the sites listed here below. Get out of your armchairs, stand behind people like Stephen, who are the real fighters in our world, and let's get the truth released, so we can get on with what it will mean to us, as a civilization, to finally recognize that we are not orphans here, on the "little blue dot." There is a greater design and we are in time to be part of it.

... Let's do it before something bad happens to interrupt this whole process.

■ *Let's do this for a new horizon for humanity and for the planet at large!*

We're right on the threshold. Thank you, Patricia, for your help in getting this out to your audience.

■ ■ ■ ■ ■

Stephen Bassett is the executive director of Paradigm Research Group and the political action committee X-PPAC. Contact him at:
    PRG@paradigmresearchgroup.org

Web sites:
    www.paradigmresearchgroup.org
    www.x-ppac.org
    www.faxonwashington.org
    www.x-conference.com

The Crop Circle Enigma

Andy Thomas

A leading researcher of unexplained mysteries, Andy Thomas is probably the world's most prolific writer on crop circles. Of his many books, *Vital Signs* is widely considered the definitive guide to the circle phenomenon. His latest title, *The Truth Agenda,* explores the link between paranormal mysteries and global cover-ups. Andy is founder of Changing Times, which holds events on political truth and liberty issues, and also directs Vital Signs Publishing, producing books on "the signs of our times." He is a keynote speaker in England and abroad, and has made numerous radio and television appearances. Without a doubt, he is a treasure chest of field experience and the objective mind, joined together to bring the story of the unexplained phenomenon of our days to the greater public.

■ *This is sure to be one of my most daring of conversations—the heated debate over the crop circle phenomenon. From spiritualists to skeptics and everyone in between, the question over crop circles simply doesn't go away. You are the perfect person to bring us enlightenment upon this controversial subject, Andy, since you complete the triangle, the balancing of the poles between believers and nonbelievers, bringing your blend of science and spirit to help us understand the key elements of this phenomenon.*

Can you give us a brief overview about just what we are talking about here? Many people still do not really know about crop circles and their appearance in this time of human evolution. Others will have their own opinions. We look to you to give us the informed insider's perspective.

Huge, complex, and beautiful patterns have been found swirled into crop fields around the world every year for several decades now, but there are reports of formations going back to the 1600s, and photos from as early as the 1930s, so this is not just a recent development—even though there has clearly been a major leap in complexity since 1990.

Their origin and purpose remain a complete mystery. Despite attempts to dismiss them as the work of human artists, some believe the evidence points to a much stranger explanation. Wherever they come from, these spectacular designs often display very clear symbolism—scientific, esoteric, and astronomical—and demonstrate some extraordinary, unexplained effects.

As you know, there has been intense debate over the circles' origins. Some believe they are communications from extraterrestrials, pointing to the many sightings and videos of aerial phenomena seen in connection with crop formations. Others feel the lights may be properties of an unknown natural energy, which produces complex ground patterns. Others, still, have cited everything from Mother Earth to nature spirits as being responsible.

Experiments with the power of the mind have suggested it is possible to influence the creation of certain shapes, leading some to believe psychic forces are involved. Beyond this, most other popular explanations for the crop circles have revolved around human activity, either involving satellite technology or, more usually, the simple actions of pranksters and landscape artists. However, in demonstrations, human teams have either struggled to reproduce designs as geometrically complex as many seen in the fields or taken long hours to produce anything even approaching them.

Certain formations have appeared within very short periods of time and the geometrical calculation and construction required for some simply could not be carried out in one night. Some of the patterns have shown breathtaking symbolic qualities, but most remain obscure in their meaning and are open to interpretation, seeming to reflect multicultural symbolism.

Laboratory evidence on circle-affected crop has identified biological changes taking place at a cellular level, suggesting the partial involvement of microwave energy. Other physical tests have shown anomalies not yet replicated by man-made experiments. These, together with the lights, eyewitness accounts, reports of malfunctioning electronic equipment, and health effects on people visiting circles, suggest the phenomenon should be looked at far more closely.

Whether the crop formations are warnings, messages of greeting, or abstract doodles remains to be seen, while they amaze and frustrate in equal measure. Even within the crop circle research community itself, there has been much intense debate, disagreement, and division—but also much positivism and inspiration, sparked by the deep questions raised and through the simple influence of beauty in people's lives.

■ *Many people, guided (or misguided) by the media, still believe this is a man-made phenomenon and that it is all a cosmic hoax. What do you have to say about that?*

I would have to ask how anybody could possibly know that all crop circles are man-made, because that view, I believe, can only be an opinion. Certainly, if you are looking at the evidence, there is nothing to support the total-hoax theory. Indeed, when you look at the evidence for what has appeared in the last few years, what you see is a phenomenon that simply refuses to go away. It is hard to believe that, if all formations were man-made, it would still be going on after all these years and with the level of interest that still persists in some quarters today.

I think what we have to do is to concentrate on the actual evidence—not on hearsay and opinion.

■ *Great, so let us know a little bit about the evidence that you have discovered in recent years.*

Now is an interesting time, because there is a popular perception amongst researchers on the ground that the last few years have been a bonanza for the phenomenon. In truth, the numbers and ingenuity of the designs have been remarkably consistent over the last decade. Statistics go up and down from year to year, with a little drop around 2006, but essentially the figures have been fairly steady, with around two hundred fifty or so formations reported from around the world each season. Yet, the feeling from the research community is that the phenomenon has come back with a vengeance in recent times. Certainly the quality of the patterns that have been appearing now is extremely high, and I think what is being reflected in the renewed enthusiasm is the happy fact that, after some years of general disillusionment, public opinion has turned back toward the crop circles in a more open-minded way. This is almost certainly because of the phenomenon's persistent reappearance, year upon year, despite all the cynical press it receives.

■ *Most of the crop circles seem to be centered in Wiltshire, England, attributed to many factors—including the number of sacred sites found there. Can you describe other factors that might have to do with this propensity of the crop makers (whoever they are) to create their images there?*

There has been great discussion about why England, in particular, has been blessed with so many crop formations. One factor that has been clearly identified is that they seem to cluster in areas of geological strata that hold a lot of water. In southern England, for instance, there is no question that the majority of formations appear on top of chalk. The circles follow its distribution closely, with the center being in Wiltshire, as you say. Chalk is an excellent water carrier (or aquifer), and water is known to generate natural energy effects, so many believe that this is helping to create crop patterns at energetic hotspots, through some unknown process.

It may be that, by chance or otherwise, England simply has the geology most conducive to crop circle creation, and it could well be that natural forces are at least one component in the placing of these designs, whatever other factors may be involved. Earth energies seem to have drawn the builders of ancient sacred sites to the same places, and therein probably lies the connection. But it may also be that the mindset of English people is particularly suitable for the dissemination of the crop circle "message."

The English are very good at bridging international divides, with a healthy balance of cynicism and wonder, and they don't keep things to themselves. What happens in England tends to be discussed around the world—perhaps this combination of appropriate natural energies and inquiring minds with a propensity for good global outreach is why this nation is being targeted in particular. It's something that's very hard to pin down for sure, and it is easy to forget that this is also a worldwide mystery.

■ *Are the overseas formations more simplistic, or do they also reflect the complexity of those in Wiltshire?*

In the last decade, there has been an increase in the numbers of formations in countries like Germany, for instance, which for some years was second only to England in terms of activity. More recently, Italy has come much more to the fore, and countries like Switzerland and Poland have seen an upsurge too. Certainly the formations in Germany have been very much more complex than some in other countries, and more reminiscent of the ones we have seen in England. Italy has been more inconsistent in terms of quality, but they have also had a number of very beautiful patterns as well, one or two of which you would quite happily expect to find in the fields around Avebury in England.

■ *Do you know a Web site or other source where one can go to look at these global formations? We don't usually get a glimpse of them on the mainstream Web sites. Can you give a recommendation of where we can see those?*

There are several sites that feature global formations. If you go to the www.cropcircleconnector.com Web site, which is well known as one of the key sites for crop circle reports, they contain pages about the overseas formations, so the information is easily available. Another Web site which is very good for global and English events is an Italian site called www.x-cosmos.it/cropcircles, which lists formations from all around the world in one archive list, so that you can see, day by day, what is appearing.

■ *Can you update us as to the quality of very recent crop circles?*

As far as England is concerned, 2008 continued the trend set in 2007 of an initial burst of very fine formations appearing (from as early as April onward) first in the yellow rapeseed. Then, after a short lull in the early summer, we have a string of fairly regular ambitious patterns in crops like wheat and barley, one appearing every other day or so, culminating in some huge and wonderful events in July and August. 2008 even saw a number of formations in maize (not an easy plant to flatten), going into late September. The geometrical verve and breathtaking symbolism of many of them never fails to impress.

2009 got off to a very busy start, with a seeming increase in the amount of complex rapeseed designs.

■ *Tell me, Andy, do you get into all of the formations every year?*

I don't think many people would be able to say that they got into all the formations of any one year. There are too many of them, and they are sometimes too widely scattered to be able to have the time to achieve that, much as it would be nice to. I live several hours' drive away from the main Wiltshire heartlands and have to make specific trips there when I can, but I am happy if I can manage to visit a reasonable number each year. Very special formations can stimulate trips out, though!

I have been into several hundred crop circles over the years, of all types, and feel privileged to have been able to do that. I have learned so much from that personal exploration on the ground. Any serious researcher needs to have hands-on experience if they can, not just look at aerial photographs. Certainly, there are people who do spend their whole summers in Wiltshire, either because they live there or because they base themselves there for a few months, and they spend their weeks driving around to as many of the formations as they can ... or flying over them, if they have the means.

■ *In my own experience in the crop formations, which I have been visiting for over ten years now, I found that there have been some pretty phenomenal designs that have really captured our imaginations. Are there any particular ones that you have found, in your years of experience, to be of an exceptional significance?*

For me, the astronomical designs, which I will discuss later, are some of the most impressive and important. Obviously, some of the larger, more complex designs tend to catch the eye, but sometimes it can be the little, more personal ones that can mean the most, as with the one that arrived as a result of psychic interaction experiments carried out by our Sussex team. It's a very individual thing—some people will respond very positively to particular designs, where others may feel less inspired. But that's the joy of the phenomenon: there's something there for everybody. Sooner or later, the symbolism will speak to an individual and reflect all manner of mindsets and creeds. I don't believe there is just one meaning or interpretation to any of these things.

■ *So, let me pick your brain and find out (although I know your answer is going to tend toward neutrality) just what is your personal experience of what's bringing in these circles? Can you divulge your innermost opinion?*

This is a very difficult area, because some people are very committed to their own theories of what makes crop circles. The main thrust of my dedication to the subject in the last fifteen years has been to give lectures to a lot of different groups, not just in England but also in other countries, and also to write books, both of which are attempting to make a plausible case to newcomers that something fascinating and potentially important is occurring. By sharing this phenomenon with as many diverse types of people as possible, I hope to encourage proper awareness of what is going on, and help further an expansion of our understanding. What I have found over the years is that people don't necessarily want to hear anyone's personal opinion. What they want, and need, is to hear an overview of all the possible theories, presented with the relevant evidence, so that they can draw their own conclusions.

As time has gone by, I have very much come into a center ground where I try to listen to everybody's views and be as open to all the different possibilities as I can be. Even after an involvement with crop circles for nearly twenty years now, I have to say that I don't yet see absolute evidence in any one direction to support a specific viewpoint. Given the absence of a clear answer, therefore, what I think we have to try to do is to assess some of the basic assumptions about what is going on. One of the observations that can be made is that the complexity of at least some of the designs seems to suggest some level of intelligence at work.

What level of intelligence we are talking about, and whether it is extra-dimensional, extraterrestrial, or indeed human consciousness (in a noncorporeal sense) at work is a big question. It may well be that it is a combination of all of these things. Certainly, there have been a number of experiments where people have tried to psychically interact with the phenomenon, trying to get certain patterns to appear as crop circles, by using meditation experiments and methods like that.

■ *And many people claim to have achieved it.*

That's true, and I have my own experience with that type of work, which I was involved with in the mid-1990s. Here in Sussex, which is the county in England where I live, we carried out some experiments with the medium and channel Paul Bura, together with some other very good channels. What we were originally trying to do was discover, by psychic means, where and when a new formation would appear, so that we could actually go along and witness it.

Although we never achieved that ultimate aim, we did manage to do what a number of other groups have done, which was seemingly to influence the creation of a specific crop pattern.

We divined the chosen shape we wanted to have appear as a crop circle through various means over several weeks, using psychic guidance and contemplation, and we eventually performed a meditation up on an ancient sacred site called Wolstonbury Hill, an Iron Age hill fort near Brighton, in East Sussex. We focused our thoughts on the design that we wanted and—astonishingly—on that exact night, the one we chose actually appeared as a Sussex crop formation.

Some people have been cynical about this, and we've been accused of going out there and making it ourselves, all the usual things. But we didn't, and nobody outside of our small and very close-knit group knew what shape had been chosen. Yet there it was in the field. If it was just us that had achieved this, it could be written off as coincidence, but other groups have done similar work. We wrote up our experiences into a book, *Quest for Contact,* which, rather nicely, I have heard has helped to inspire further attempts around the world.

■ *Let's assume that you have a wide audience that would believe in that. I would like to know how you worked together to fixate a certain geometrical pattern.*

We received messages through Paul, who was our main channel. He had a number of visions and directions that we should play certain notes of music. The notes we used were A, B flat, C sharp, D, F, and G sharp, which we recorded on a synthesizer using sine waves: a very pure form of sound. We did this because a number of people believe that sound is involved in the process of creating these shapes. This is not to say that sound is the whole answer to crop circles, but it may be an important component.

Through our divined notes of music and the various other directions we were given, we came to the conclusion that what we were going to receive was a combination of six circular elements. A member of our group, Barry Reynolds, drew on a piece of paper the shape he thought would result. We were delighted and shocked when a formation in this very shape appeared the day after our experiment.

■ *So you can either ask yourself was this channel telepathically linking into the consciousness that was sending the formation in? Or did you as a group interact and actually direct the communicators?*

That is the big question—which way around did it occur?

■ *I mean there is no question that there was interaction, with this kind of precision.*

My personal feelings are that what occurred in our particular case and, I suspect, with other groups who have carried out similar successful experiments is that a combination of things were at work: human consciousness interacting with something else. What that "something else" actually is, I have never been able to put my finger on, although I know that some people have some very particular views on this. I know that you, too, have very particular views on where crop circles come from, for instance, but because I try to stay completely open-minded, I find myself unable to completely commit to one point of view on what this other consciousness is.

■ *I really appreciate that position as well! As for my own viewpoint and experience, I try to absolutely give no prejudicial opinion about what's happening when I bring people into the circles, because these experiences are individual and very personal. For some people they are life-altering. I really don't want anybody else's opinion, including my own, imposed upon other free-thinking individuals.*

Yes, I think it is very important not to do that.

■ *When I went first into the Stonehenge Julia Set in 1996, I had no pre-set opinions; I went in by myself and I was given a lot of direct information. This triggered the collection of writings The Sirian Revelations, as you know. Many people who are going in are having these very, very big immediate experiences . . . it is definitely intriguing, to say the least, for so many who make it in.*

I do have to say that in 2007 especially, the circles were incredibly powerful, much more than the previous year, to be sure. But nothing has ever been as mind-altering for me as that one formation—I love to call it "my crop circle"—at Stonehenge.

As you say, the experiences people have inside these formations are very personal and very subjective, most certainly. What is so exciting and interesting is that a design that may not resonate in any way at all with one person, yet it may be the very pattern that somebody else opens up to, in an experience that may completely change their life. This happens again and again; certain formations resonate with certain individuals and not with others. Each person will find a particular design that will really work for them and that is why I feel that to load the dice, to say to people that a certain formation means only "this," or to say that the phenomenon only originates from "here," is not fair. Doing so takes away from that very personal experience that everybody should be free to experience on their own, with this mystery.

■ *Agreed. There are people that you immediately embrace and others that you just cannot connect with, so it would make sense that there is no predetermined experience about any other level of consciousness—which some, but not all, are encountering in the fields.*

That's right. It is almost as if every possible style and level of crop pattern is being used, as if the phenomenon is designed to reach out to as many different people as possible. If one design doesn't "speak" to you, another will. The wonderful thing about the crop circle mystery is that it is very universal in what it embraces. It doesn't exclude anybody; every level of opinion can come into the debate, including not only speculation about the nature of the consciousness at work, but also about the mechanism behind the formations and whether it involves natural, dowsable earth energies or even weather-related phenomena. Everything can be a legitimate part of the discussion, and indeed it may all connect up somewhere in the middle. Nobody is excluded from the party, as it were, and the mystery gathers people together as a result.

■ *When entering into the circles, it is quite amazing how everyone respects the others' space; people are usually quite silent; there is so much reverence and humility in these moments ... and that in itself is a remarkable thing to be part of.*

And in between the reverence ... people talk to each other! This is a marvelous thing. You walk into a crop circle and you meet people you have never seen before in your life. If you saw them out on the street or at a bus stop you probably wouldn't even look twice at them, but, in a crop circle, people tend to communicate with each other. You'll invariably meet somebody who has come from the other side of the world to look at the formations, or they will unexpectedly share an experience that, to you, will prove valuable to hear.

There is something special about these patterns; they provide gathering spots and they open up channels of communication between people, which is wonderful to see. For those who have never been to a crop circle, it is sometimes difficult to understand how it can be like that, but it is. You enter a field and suddenly you are standing in what feels like sacred space. It does very much have that feeling of being in a cathedral or, as a number of people have observed, a temple. Because of this, you suddenly find that your behavior changes and, when you meet other people going through the same experience, suddenly you have something in common. You open up to them and communication occurs. If that is all the formations have ever done, that, in itself, has been a very important gift.

■ *It is quite remarkable that so very few people are actually getting into the formations. As important a phenomenon as it is, I am always surprised to see that often it is the same people going in and out, year after year, that I am running into. I just can't believe, despite the media and the hype and the skeptical journalism, that there aren't more people flocking to these circles, which I personally believe are some of the most important things happening on the planet.*

It is an interesting thing that if you go to one of the crop circle Web sites that has a hit counter, like Crop Circle Connector, you can see how many Web surfers have looked at each formation. The amount of people who are keeping an eye on these things is actually shockingly low, when you consider that there is a whole planet of human beings out there. Even an extraordinary formation may only have something like ten thousand hits on it. That is "hits," not individuals, and people often go back to the pages again and again to look at them. That may only be two to three thousand people at the most, out of the whole world, who are looking at these things.

It's shocking, but true. I think that those in the crop circle community often have a slightly overinflated opinion of how big the interest is. The majority of the world's population knows nothing about this phenomenon, incredibly, and it is a very small group of people that is monitoring it and interacting with it on a regular basis. So, yes, it may turn out to be one of the most important things happening, but those acting as a conduit to show the formations to the rest of the world are remarkably few.

Because of this, I think there is a great responsibility for those conduits to report what is happening in the fields very openly and without bias. For example, everyone accepts that a proportion of crop circles is man-made, but it can be irritating to hear some alleged researchers saying, "They're all man-made now" or "This circle is man-made and I can tell just by looking at a photo." Not only does this prejudice people on the outside, who will simply conclude that perhaps there is no phenomenon to be interested in after all, but I also do not see how people can know for sure what is man-made unless they happened to be there when a formation was manually created, and have incontrovertible evidence.

Everyone is entitled to an opinion, as long as it is clearly described as such. This is where we do have to be very responsible because, as so few people deal with the crop circles on a one-to-one basis, what the active researchers have to say affects the level of general interest.

As you say: "Why aren't there more people in these formations?"

I would argue that this is because most are just not made properly aware of the remarkable evidence that proves there **is** something there to be interested in! This is why I spend so much time providing outreach on the subject, to take it beyond the narrow borders of the circle community and its parochial concerns, and to try to get new people excited. The disillusionment of a few disaffected researchers has been allowed to infect the line of communication to the outside world, which is already skeptical enough, and that needs to change.

The farmers, of course, are very happy that there aren't more people out there, but considering how important this phenomenon could be, the small scale of interest is indeed strange. But as I said earlier, the persistence of the phenomenon does keep a little intrigue alive in the minds of the public. I know, from my own experiences with audiences, that minds can easily change and enthusiasm can be fired, when they are presented with sensible evidence in a rational way.

■ *Using the left-brain, analytical mind, when looking at a diagram or pictures of the circles, can suppress any interaction, intuitive or psychic, of the right-brain functions that control the experience of the subtle realms. So, I do find it difficult to understand how people can feel comfortable claiming to know everything about a formation, simply by dissecting an image or diagram of a formation.*

They're looking, they're analyzing the message, the diagram, the art, the specifics of the geometry, but they are not necessarily feeling it. I think that one of the most exciting aspects of the phenomenon is that we seem to be in the middle of something unknown to our human experience. Or perhaps it is just dormant in our ancestral memory. I do believe that we need to move beyond our mental bodies, however valid it is to use that analytical process, and let ourselves experience the ebb and flow of the energies present.

I think that this is another important point. If you go into the crop circle subject wanting only to analyze it technically, you can only go so far with such an approach. This is not to say that it shouldn't happen at all, because there has been a lot of very valuable insight gained by studying the geometry, for instance.

There are also biological anomalies occurring within the flattened plants, and some other fascinating science that doesn't get the attention it deserves, and this must, and should, be investigated. But one also needs to stand back and give in to intuition as well. If you don't, then sooner or later you will become frustrated with the phenomenon.

I am afraid that crop circle research is littered with people who were very sure they were going to prove "x" theory or "y" theory about the mystery, but then couldn't—because the circles don't seem to respond to that kind of one-track scrutiny. These tend to become the skeptical types, who get very frustrated and come across very cynically, because they don't get from the phenomenon what they want intellectually. This is often when they throw up their hands and say, "They must all be man-made, then!" when what they are really expressing is a personal frustration that they were not able to go further with it, at their own level.

Intellectual approaches come and go, but the phenomenon continues on. It does seem to be the people that can open to it on a heart level who stay the course. I think there is an interesting observation in that.

■ *Well, it is quite a feat to silence the chattering mind of the logic that wants to control, control, control. But to everyone who is listening, I can't tell you enough how important it is to get yourself to these circles. I am really not all that up to date on the global phenomenon, but I do go to England almost every summer and if you can make it there ... do it! It is a very exciting, very deeply enhancing, and, I believe, very* important *thing happening on our planet.*

Amazingly, because the media never wants to share the really important events, the public is kept focused on Paris Hilton and Britney Spears and all the insignificant goings-on of the superficial layers ... and war of course. Always war.

Here is this remarkable phenomenon, occurring constantly, year after year, possibly one of the first clear signs of extraterrestrial or extra-dimensional communication! It deems investigation at the personal and global levels.

Agreed. It is an experience which you can't really replace with anything else, and with such a concentration of designs in the Avebury area, you can go there some years, drive around a five-mile radius, and see on virtually every horizon another formation, albeit obliquely. Each year, these mysteries are being thrown down all around there and the feeling of being surrounded by it is really quite something. So, yes, for anybody who has not had the experience, it is worth taking a vacation in England and spending some time in the Avebury area.

■ *I agree and I do want to say that every year I do bring a group through my SoulQuest™ Journeys program, to help awaken people to this extraordinary phenomenon. If you are interested in meeting with Andy, we can always try to arrange that.*

Are you doing the Glastonbury Symposium again this year?

Yes, I am one of the organizers. The symposium is an annual event that has been running since 1990, through which we explore crop circles and many other related mysteries. It is a fascinating and rewarding experience to gather together with like-minded individuals searching for deeper truths about their world, and Glastonbury, in itself, is a beautiful place to be.

■ *I would like to ask you, in all of your experience in the circles, have you had any light phenomena or sightings while you were inside, besides military helicopters?*

I have indeed witnessed unexplained lights. I wasn't inside a formation at the time, but back in 1993 a group of colleagues and I were at Alton Barnes in Wiltshire, and there were two evenings when we witnessed strange phenomena. We were looking down over the East Field, as it is called, which is a vast area where many famous crop designs have appeared, and there was a formation nearby.

We all saw floating, pale, luminescent globes of light flying over the fields and we were able to verify with each other what we were seeing. Then, two nights later, we were parked up at the bottom of the East Field, on a level with the field, looking back up toward a place called "Knap Hill." In the very early hours of the morning, we witnessed a series of very large, dark flickering shapes in the sky.

To this day I cannot really say what they were, but they were very odd. They weren't craft; it was almost like something from some other dimension, a strange, flickering "unreality" before our eyes. It was very eerie, although I would like to witness it again, as it would be valuable to reapproach it with the knowledge that I have gained in the years since.

That was my main experience of such things, and I have heard of many people who have witnessed the lights or other similar strange shapes. This is another whole aspect of the circle mystery—the peripheral, so-called "paranormal phenomena" that go on around it. The balls of light, especially, have been witnessed by many people and videoed on several occasions over the years.

■ *Wasn't it in 2007 that there was a video of a ball of light in an East Field formation?*

The large 2007 East Field formation became central to some heated debates. It was a huge formation, one thousand thirty-three feet in total length, with many, many circles spread across the fields. Three researchers were up on the hill the night it appeared—the same hill, in fact, from where I witnessed my light phenomena, years before. They were filming the field and they witnessed a very bright flash in the sky at about 3:00 a.m.

The story goes that when they analyzed the footage, the flash showed up as an interference blip on their video. With enhancement of the film, it has been claimed that the formation is not there at about 1:45 a.m., yet at about 3:20 a.m. you can clearly see the formation. This is potentially important information, because manmade formations can take many hours to make, and something of this scale couldn't possibly have been constructed in such a short time span.

The flash may have been something to do with the formation's appearance. As ever, of course, arguments over the details and controversial claims from skeptics created the usual crop circle situation of not being able to get to the absolute truth of what occurred. It is very difficult to reach ultimate conclusions in these situations—and it always has been. I have never known a year without a controversy like this, and it does seem to be that paradoxical nature again, that the phenomenon never quite lets you get to the truth of anything.

That almost seems to be part of its purpose. Perhaps, without the mystery, there would be nothing to explore in our minds, and thus no expansion of consciousness stimulated by the journey of the questioning. So it always makes sure that there is never irrefutable evidence of anything! It is interesting to note, though, that over the same East Field formation there were many alleged sightings of the military helicopters you mention.

■ *I have never been in a crop circle without military helicopters overhead … not one!*

It should be remembered that the helicopters are an entirely normal part of life in the Vale of Pewsey, where Alton Barnes lies. It is an area not far from the military bases of Salisbury Plain and they do train helicopter pilots there. The pilots use the crop formations as useful markers to hover around and train with, so it's not always evidence of anything sinister. That said … there have been some instances where it seems the military have shown more interest than usual.

■ *I was in a crop circle one year and there were two or three photographers with tripods attempting to photograph the formation, but all their cameras malfunctioned. A helicopter flew overhead and there was a military person sticking out of the helicopter taking pictures of us. I thought of it as a touch of irony … that we were curious about the circles, and they were curious about the curious.*

Some people believe sightings like this prove that the military know what's going on, or that it is actually responsible for making the crop circles. This is a presumption, though. I am sure the military is interested, but that doesn't mean that its investigators know any more than the next person.

■ *I suppose there may be a possibility that they are as bewildered as we are and unable to control it. They are, after all, concerned with "controlling the borders," right?*

I think we have to be careful not to go too much into the realms of paranoia. Maybe the helicopter pilots are just playing dark games, having a bit of fun at people's expense. Certainly some of them are irresponsible to come down so low over people's heads. It could be that they are running strange tests of their own, but we just can't say for sure.

As for the oft-spoken theory that the military actually makes crop circles through some secret technology, this cannot be entirely ruled out, but equally there is no evidence for it. It is puzzling to me why the military would spend its days making crop formations so publicly, and then spend so much time dismissing them, as it seems to have done in the past. Why not test hidden technology on restricted land, instead of making circles in very public places? It doesn't seem to make any sense, unless there is some bizarre double bluff going on. These arguments will run on and on.

■ *Anything is possible—I do agree with you on that one. That is one of the most exciting things about this whole experience. If you can open your mind to any number of possibilities without paranoia, but certainly observant of all the unfolding activities, you find you are, in many ways, in some kind of space-time warp, exploring the unknown.*

The point that I have made over the years, in my lectures and books, is that the element of mystery in the crop circles is maybe one of the most important things. Because this phenomenon doesn't serve up the answers on a plate, it always leaves us in the middle, in that state of "unknowing," as you say.

We have to reach within ourselves to find the answer.

It seems to me, in observing the people that I have interacted with over the years, that this is how the circles stimulate an evolution of consciousness—in the questioning, in the constant search to find out what these things mean and where they come from. In the end, you just have to **feel** it, as much as applying your mind to **understanding** it. It is in not knowing where the circles come from that an inner awakening evolves.

It may well be that the whole thing was never designed to give us just one message or one form of information, which some people are searching for. Perhaps the phenomenon is there purely to stimulate, to deliberately stir up controversy and debate, because that's how evolution arises.

Certainly, if we had formations with writing next to them in the fields, telling us where they came from, what made them, and how they were made, it would instantly deflate a lot of the mystique. Not knowing is one of the very things that makes them so exciting. It creates that "inclusivity" that we were speaking of earlier on. It means that no one person can claim exclusive rights over an explanation and say, "It's definitely this" or "It's definitely that." There is simply no way of being sure at this juncture.

Our understanding of crop circles, for all the intriguing speculation and the many valuable experiments, is still a completely blank sheet—and yet, at the other end of this is beauty: extreme examples of wonderful geometry and fantastic patterns that are very much there and that seem to change people's lives. I think this is a key point.

Most unexplained phenomena don't last for very long. With UFOs, for example, they come and they go and one has to be in the right place at the right time to see them. Ghosts are similar in that respect. But with crop circles, they are very definitely there, and sometimes remain there for quite some time. You can take people by the hand and say, "Come and see this!"

They can touch the formations, they can smell them, and walk around and inside them. This is very unusual in the mystery realms.

■ *One of the most powerful formations I have ever been in was the 2007 Sugar Hill formation, which was only up for a couple of days, I hear, and they mowed it down. Did you get into that one?*

I went to see the Sugar Hill formation, but missed it in its prime. It had been there in the morning, but by the time we arrived, it had been chopped down. We went and stood in it anyway. Even when formations have been harvested, you can actually see quite a lot left, if the blades of the harvesting machines weren't able to be lowered enough to pick up all of the flattened crop.

The vastness of some of these designs is extraordinary; Sugar Hill was a case in point. It was several hundred feet in diameter and had thirty-six triangles around the outside, and then another smaller thirty-six triangles in its inner rim. Inside these were the optical illusions of many cubes, with very three-dimensional qualities.

Even after all this evolution in design, much of the public still thinks crop circles are just basic circles, and they have no idea that things like this are appearing. The formations deserve a better name, really, but "crop circle" has stuck in the mainstream.

Whichever way you look at this phenomenon, it demands some respect, which it rarely gets—particularly from the media, which still laugh or scoff at them. Yet, you only have to look at some of the patterns sensibly for a few moments to realize that something very special is going on. If one tries drawing some of these patterns just on a piece of paper, it soon becomes clear what extraordinary capability it would take to lay them out so accurately and consistently in an open field. Yet, these are appearing regularly, on huge scales, sometimes within very short periods of time.

If just rendering diagrams like this can take so long, we have to consider the conditions under which they are arriving. These designs are clearly made with no practice-runs; otherwise, we would find the "tryouts" too ... but we don't. Whatever is at work in the fields is getting these works of genius right—first go! Errors can't be rubbed out easily in a field; once crop is flat, it's flat, and can't be made to stand again.

Out of the thousands of formations we have had over the years, so many of them are works of precision. Were they all made by humans, as our media insist, we would surely spot human error more than we do.

■ *Can you talk about the pictogram that came down in 2001 at the Chilbolton radio telescope? It was a completely new experience. Instead of having laid crops, there were just these tufts of crops forming an almost digital image.*

The Chilbolton radio telescope is used for atmospheric and weather research. Next to it, two glyphs appeared a few days from each other. One was a curiously altered version of the famous signal that the SETI project beamed out into space in 1974. Our version described human life on Earth, but the crop "reply" seemed to depict another life form altogether. The accompanying formation was the rendering of a face. The face was made out of dots (or pixels) made from little pillars of crop.

On the ground, people had no idea as to what it was—only aerial shots revealed it as a face. You have to blur the face to really appreciate it—when this occurs, a very three-dimensional humanoid visage is revealed. Obviously, this caused a lot of deep speculation. Was it an alien face? Was it the claimed face on Mars? Others said it reminded them of the Turin Shroud—the face of Christ. Others said it was Tony Blair! More recently, some have said it resembles Barack Obama. You have your pick of possibilities!

The year after, we had one more formation that seemed to pick up on a similar theme to the Chilbolton events.

■ *This was the ET-disk formation that came down in August 2002. As far as hoaxes go, it sounds like a pretty complicated glyph to create!*

This appeared just outside of Winchester in Hampshire and looked like an archetypal extraterrestrial holding a disk of binary data. It was extraordinary. Instead of the picture of the extraterrestrial being rendered in dots or pixels, it was formed from lines, which scanned across, getting wider, getting thinner, gradually making up the pattern. It was very controversial, but a work of genius in terms of its construction.

The disk had an encoded message in it—the only one of its kind ever received in a crop formation. The disk, full of dots and dashes, is actually binary code and can be read using what is known as the "ASCII system." ASCII can be used to generate text on a computer. When this disk was decoded, it turned out to be a warning, albeit with a note of hope. The message reads (exactly as written here):

> *Beware the bearers of FALSE gifts & their BROKEN PROMISES.*
> *Much PAIN but still time. Believe. There is GOOD out there.*
> *We oppose DECEPTION. Conduit CLOSING.*

A little computer bell sound is even signified in the coding at the end. This message obviously caused great debate. As it is plainly a warning, who or what were "they" warning us about? Was it a warning from beings that look like the image itself? Or was it a warning about them? Many thought it might even be a warning about our own governments, and the shadowy behind-the-scenes rulers many believe exist. Either way, this was one of the most controversial and yet most amazing formations we have ever had.

■ *Right, because whether or not you are focused on who the messenger is referring to, nonetheless you cannot help but ask yourself: "How in the world could these precise binary codes have been splayed into a field of wheat?"*

I spent some time inside that formation and from the center of the disk to the outside it took a good twenty to twenty-five minutes to walk, because it was a spiral. A lot of people think it was just concentric rings, but, in fact, it was a spiral—like a vinyl record album. It was extraordinary in scale and, as you say, they only had to mess up a couple of the blocks and the coding would have been corrupted. I think there was one word, "believe," that was slightly damaged by the wind and wasn't so clear to decipher, but most likely "believe" is the word.

This is the remarkable level of sophistication that we are dealing with in the fields. Years on now, and that formation has never been claimed or explained away by any of the usual skeptics. It's still a major mystery.

■ *That is a revealing thing about the human so-called "circle makers"— they don't claim a lot of these circles, have you noticed?*

Obviously, their argument is that if they actively claim them then they will, of course, be prosecuted—but that leaves them in the rather comfortable position of never having to justify their many insinuations. And that is what they do—they merely insinuate authorship of formations, so no clarity is ever reached on man-made designs, because there is no more evidence for human claims than there is for the paranormal explanations.

At the end of the day, the patterns created for television programs soon reveal the limitations of human circle makers, and the many hours their works take to make mean that a significant proportion of crop formations simply cannot be explained that way. Nobody would deny that there are some man-made formations, but put together with the statistical, geometrical, and biological evidence (not to mention the eyewitness accounts of people watching crop circles take form), it soon becomes clear that there is a phenomenon at work that humans can only partly explain.

■ *Even with the people who are going in and creating some formations, and I am sure there are some, even there is seen the reflection of the brilliance and the wonder and the experience! They are communicating through geometry, with sacred form. Whether or not they realize it, their very actions are more than just their attempt to debunk the experience. They are responding in some way to the wonder of what is going on.*

This is probably true in some cases, but I also think there is a certain level of mischief that goes on, that is not so well intentioned. Certainly, some of the people who claim to make formations have been very dark and unpleasant in their dealings with circle researchers over the years, which seems to me to sit uncomfortably with the suggestion that they are doing it for spiritual purposes. Nonetheless, it could be argued that any crop formation that appears goes into the general mix, and it still draws awareness to the fact that something interesting is going on in the fields.

■ *Are there any other formations that really stand out in your mind, with regard to the quality of the formation itself and/or the apparently galactic messages they seem to represent?*

I still believe that the astronomical designs are some of the most intriguing and potentially important occurrences. In 1994, for instance, three galaxy-shaped crop glyphs displayed a conjunction of planets in the star constellation Cetus—as it would have been on the 6th of April 2000. When that night finally arrived, an unusually large solar flare caused auroral displays right across the globe, to areas that wouldn't normally get them. Put together with the fact that a number of other astronomical designs have shown dates and times which, when those dates and times have then arrived, have coincided with unusual—and unpredictable—celestial phenomena, the astonishing implication is that we are dealing with a force that seems to have the power of prediction.

The odds that the dates indicated in the formations were just "coincidence" are really unbelievably high.

Many people think that the crop circles are trying to call our attention to the fact that something big is about to occur in our solar system. Our sun, in particular, has been behaving rather erratically in recent years, breaking records for sizes of solar flares and with a sunspot cycle that seems out of kilter at the moment. NASA is predicting that the peak of the current eleven-year cycle could be the most active we have known in centuries, occurring somewhere around 2012–2013.

This would tie in with the many prophecies, ancient and modern, that seem to point to the year 2012 as being a huge time of upheaval and transformation. Some believe that the crop circles are intrinsically tied up with calling our attention to 2012, and the awakening process many think is now occurring. It is certainly the case that direct references to the ancient calendars and the 2012 cycle have now been incorporated into several different crop circles.

In 2008, a formation at Avebury Manor depicted an almost-accurate diagram of our solar system as it will look on December 21, 2012—believed by most key 2012 researchers to be the exact day the Mayan calendar ends and begins again. Only Pluto seems to be in an unexpected position, which has caused some debate.

The idea that the crop circles are heralds of change, calling attention to a new era for our world and humankind, is a popular one... which some of the symbolism seems to support. Certainly, our response to the formations and their inherent mystery, as we have discussed, has already opened many people up to a whole new way of looking at the world, which may be a small component of bringing about a self-fulfilling prophecy of conscious evolution for 2012.

If all that this phenomenon has done is to make people aspire to higher things, and brought mystery and wonder back into their lives, then that is a major achievement in itself, whatever the future brings.

■ ■ ■ ■ ■

Andy Thomas has authored:

Fields of Mystery: The Crop Circle Phenomenon in Sussex

Quest for Contact: A True Story of Crop Circles, Psychics and UFOs (co-authored with Paul Bura)

Vital Signs: A Complete Guide to the Crop Circle Mystery

Swirled Harvest: Views from the Crop Circle Frontline

A Oneness of Mind: The Power of Collective Thought and Signs of Our Times

An Introduction to Crop Circles

The Truth Agenda: Making Sense of Unexplained Mysteries, Global Cover-Ups and 2012 Prophecies

Web sites:

www.vitalsignspublishing.co.uk

www.swirlednews.com

www.changingtimes.org.uk

Information about the annual Glastonbury Symposium can be found at: www.glastonburysymposium.co.uk.

The Spiritual Brain

■ ■ ■ ■ ■

Dr. Mario Beauregard

Dr. Mario Beauregard's groundbreaking work on the neurobiology of mystical experience at the University of Montreal has received international media acclaim. Before becoming a faculty member there, he conducted post-doctoral research at the University of Texas and the Montreal Neurological Institute (McGill University). Because of his research into the neuroscience of consciousness, he was selected by the World Media Net to be among the "One Hundred Pioneers of the 21st Century."

■ ■ ■ ■ ■

■ *Dr. Beauregard, I am always excited to experience the merging of science and spirit and you have made a remarkable effort to help us to understand more how the two co-exist and rather than contradict each other, they merge together, through your research and your personal perspectives. I would like to ask you to start this dialogue by giving us an overview into your book* The Spiritual Brain *(co-authored with Denyse O'Leary) and what underlying predominant theme defines this rather immense subject: the human mind, the human brain.*

In the field of neuroscience, the mainstream view is that mind consciousness and spirituality can be reduced totally to electrical and chemical processes in the brain—a very materialistic and "reductionistic" view. I decided to challenge this view, because there are several lines of evidence indicating that it is not correct. I have explored various lines of evidence, starting with findings from the studies we have done with the Carmelite nuns, but I also decided to present other lines of evidence.

In the kind of work we are doing we are using neuroimaging techniques like "functional magnetic presence imaging"—it is a form of scanner and it allows us to visualize which parts of the brain are active during all sorts of tests.

■ *Would that be similar to a CAT scan?*

Yes, but we are measuring the changes in terms of oxygen consumption in the blood circulating across the brain, and this phenomenon is coupled to electrical activity in the nerve cells. We also use another technique, called the EEG (electroencephalography), which measures electrical activity of the brain. We have access to all sorts of neuroimaging techniques and we measure what we call "correlates." These are simply biophysical measures of what is going on in the brain during all sorts of tests. It could be an emotional test; it could be also a spiritual state or experience.

■ *What kind of tests do you do? When you refer to your experiments as tests of "emotional" or "spiritual states"—are you evoking a thought or an emotion in the individual?*

Oh, yes. If we study emotion, for instance, we can use emotionally laden pictures, static pictures, but we also utilize film excerpts to induce a specific emotional reaction. In the case of spirituality and spiritual experiences, we asked the nuns to enter into a state of contemplation and prayer, because we were interested to see what was going on in the brain during a mystical state—a state of union with God (according to the terminology of the Carmelite nuns).

The Carmelites are members of the Roman Catholic order and this order has been oriented toward mysticism, especially by Saint Teresa of Ávila and Saint John of the Cross, during the sixteenth century. So, we did that and, effectively, we have been able to measure all sorts of very interesting changes in the brain during this state.

We were testing a specific hypothesis: namely, the "God spot" in the brain. About thirty years ago, some neurologists reported that, when epileptic seizures are located within the portion of the brain that we call the "temporal lobe," the patient would sometimes report a "spiritual experience." These neurologists concluded that the temporal lobe was the "God spot" or the "God module." But, truthfully, this interpretation was based on a very materialistic explanation of this phenomenon. In essence, they were trying to reduce spiritual experiences to simply the misfiring of neurons in this region of the brain.

■ *In fact, it is very interesting that you brought up epilepsy, because in many ancient civilizations and indigenous tribes, people with epilepsy were always revered as the real "seers."*

That's right, because it seems that, during these seizures, there is a kind of chemical and electrical shift in the brain, and this shift allows the individual to access another form of reality. In our case, we saw activation in the temporal lobe. It was there, but there were also many other brain regions activated during the mystical state, in particular regions of the brain normally associated with emotional states and also self-awareness, because there is also a marked change in terms of self-awareness during a mystical experience.

In this kind of research into what I referred to as "correlates," there is no causality involved. In other words, we do not know if the activity that we measure really relates to a reality external to the subjects of our experimentation, or whether perhaps it is simply a product of their imagination. To be able to solve this question, we have to turn to the "near-death" experience, which is a very interesting model for neuroscience, especially in the case of near-death experiences that are induced by cardiac arrest.

In the case of cardiac arrest, the brain will shut down after a short time frame of about ten to fifteen seconds, because there is no blood circulating anymore. It is very interesting, because in certain surgical procedures, the surgeon will actually induce a cardiac arrest to be able, for example, to remove giant aneurisms at the base of the brain in a region that is called the "brain stem." The brain stem controls all the vital functions such as heart rhythm and also breathing.

■ *What is the logic behind their inducing of the cardiac arrest? I imagine it is an extremely dangerous decision to make in the surgical environment.*

Cardiac arrest can be induced in order to remove all the tension in the blood system. This will allow all the blood to be drained from the head, and the body temperature to decrease to about fifty-eight degrees Fahrenheit. Basically, you put the person into clinical death for a certain period of time. This period normally ranges from between fifteen to sixty minutes; during that time, the person is clinically dead and the brain is not functioning. The likelihood of a successful surgery is actually much greater in these cases.

About fifteen years ago, there was a case of a woman named Pam Reynolds who was suffering from just such a major aneurism at the base of her brain, and her life was threatened, since an aneurism can burst at any moment. She heard of a neurosurgeon in Phoenix, Arizona, named Dr. Robert Spetzler. About thirty years ago, Dr. Spetzler devised this so-called "brain stem procedure," during which the heart is stopped and the blood is drained from the head.

In the case of Pam Reynolds, she was clinically dead for about sixty minutes and that is really quite a long time. What is fascinating is that, while she was clinically dead and her brain was not functioning anymore, she had the impression of leaving her physical body and floating above it during the surgery. She was able to draw, very accurately, some of the surgical instruments that were used during the surgery, and she was also able to report, with accuracy, the dialogue exchanged between the neurosurgeon, the cardiologist, and also the nurses who were there.

At another point during this experience, she had the sensation of floating along a tunnel at a very high speed, and at the end of the tunnel she met with deceased relatives and friends. In the last part of her experience, she met with a beautiful being of light and her life was totally transformed by this experience. This was the first case that we have where we know **for sure** that there was no physical brain activity and yet the person was able to perceive, to have consciousness, and to have emotions. Now we have other cases like this one. This case directly challenged the mainstream materialists' views in neuroscience that mind, consciousness, and spirit are simply byproducts of electrical and chemical processes in the brain.

■ *In fact, so many people are reporting on these experiences. My own mother had surgery and they nicked an artery. They had to rip her open from the throat down to her navel in emergency open-heart surgery.*

When I rushed in from Italy to see her, while she was still in intensive care, she told me she heard them say, "This one's gone. . . . Let's pull the plug," and that she was screaming from inside: "No, no, no, I'm alive, I'm here, don't abandon me!" while they were getting ready to pull the proverbial plug on her life.

It is so important that doctors understand this state of consciousness . . . so many people most surely die unnecessarily, because the doctors believe they are "gone." I have a little bit of a firsthand experience here, or at least secondhand, through this terrifying experience of my own mother.

These experiences are quite common, because the techniques of reanimation are much better now than they were thirty or fifty years ago. There are millions of cases out there, but, in terms of neuro-science research, the cardiac arrest paradigm is very interesting because, like I said before, no more than ten or fifteen seconds after cardiac arrest, the brain will shut down. Mainstream neuroscience states that there cannot be any consciousness or mind processes going on if there is no brain activity, and yet we have this proof. So it is very powerful for us researchers.

■ *Yes, I suppose we are talking about the difference between the physical brain and the mind.*

Right. It is as if the brain is acting like a kind of transducer of mind, consciousness, and spirit at another level.

■ *So, can you define the difference between mind and brain?*

Well, the brain seems to act like a sort of interface for the mind, so the mind, of course, is conditioned by the brain in terms of expression and experience in a healthy person. If there is a lesion, for instance, in specific parts of the brain, then the conscious experience and mind processes will be altered, such as in the case of Alzheimer's disease, and you have all sorts of diseases like Parkinson's disease. You have very specific destruction of nerve cells in those cases, and the destruction of the nerve cells will significantly change perception, cognition, and emotion in the people suffering from those diseases.

But, in my view, it is the relationship between mind and brain that is altered. If you are playing with the electronics at the base of your television set and if you twitch something in the components of your TV, you alter the reception of the information, but yet, the television station keeps emitting the same kind of programs, just that the reception is distorted. I think we can use this analogy to describe the mind-brain relationship.

■ *You talk a lot in the book about the God gene. This is a central theme and I am very fascinated by the discussion of it. I know how difficult it is to shrink big concepts into such short conversations, but can you give us your insights into this concept?*

I will try to simplify the whole matter. A few years ago, there was a molecular geneticist at the National Institutes of Health in the United States by the name of Dean Hamer, who wrote the book *The God Gene*. The book is based on a series of works that he did in the last decade. What he discovered, essentially, is that there is a relationship (or a correlation, technically) between a specific gene in the body and also certain traits, as measured by a scale which explores various aspects of spirituality. So there are various components in the scale that are measured, like self-transcendence and the sense of interconnection with environment. He discovered that there is a statistical relationship between a specific gene and how this gene codes for various chemical messengers in the brain.

We call these messengers "monoamines"; they refer specifically to serotonin. Serotonin is a chemical messenger that is involved in mood

regulation, but it is also involved in spiritual experiences and spirituality. There is also dopamine, which is actively involved in reward
and motivation; there is also another transmitter called "noradren-
alin," which is involved in arousal and stress. These chemical messengers are related to the components of the scale (that measures
various aspects related to spirituality) that I was talking about earlier, devised by Robert Cloninger, a psychiatrist at Washington University in St. Louis.

The problem with this very reductionistic interpretation is that
the gene in question simply referred to chemical messengers that
are involved in all sorts of functions, in addition to spirituality. We
know that, for every behavioral function, whether it is an emotion
or a condition (for instance, memory perception), it is not only one
gene selectively that will be involved, but rather thousands of genes
. . . and also the interaction between these genes. Dean Hamer has
been accused of being very reductionistic regarding the question of
spirituality and the brain. A few years after he had written his book,
he recognized publicly that perhaps he was too radical in his opinion. He agreed with some of his critics that, of course, thousands of
genes are very likely involved in what we call "spirituality" and also
"spiritual experiences."

That is the viewpoint we adopt in *The Spiritual Brain*.

■ *Would that include some of the DNA that is considered junk DNA?*

No, in his case he specifically identified a very selective gene, but he
did not refer to junk DNA. Junk DNA must also be involved.

■ *Because we know that it is not junk, we just don't know what it is doing,
right?*

Yes, exactly, so it must also be involved with what we call "spirituality."

■ *I think that it is very exciting to know that, as more and more discoveries are made about our gene codes, we are realizing that we have the potential to uncover all of this so-called "junk DNA" and learn, bit by bit, what it serves. We know nothing could be junk in the body, so that terminology just doesn't make sense at all.*

I agree.

■ *Can you tell us more then about your tests with the Carmelite nuns and what you learned from them in those tests?*

Yes, as I said before, we were testing the so-called "God module" or "God spot" hypothesis and temporal lobe. We wanted to know if this region was really the sole region involved in mystical experience. The results of our studies, both with the fMRI (functional Magnetic Resonance Imaging) technique and also with the EEG, indicate that the temporal lobe seems to be involved, this is true, but it is not the **only** region of the brain involved. In reality, these mystical states seem to be supported by a complex network of brain regions. This involves several parts of the brain that are normally associated with various aspects, like self-consciousness or self-awareness and also positive emotion, because there is very often a positive emotional component involved in a mystical experience.

There is also a brain region called the "parietal cortex," which is involved, and this region is normally associated with what we call "body representation." This is the representation made at the brain level, regarding the location of the physical body within space. During deep mystical and meditative states, this representation is altered and this brain region—the parietal cortex—of course is modified in terms of its activity, and so we saw marked changes in this region of the brain.

There has been a study done by Andrew Newberg at the University of Pennsylvania a few years ago, with people who are experienced in Buddhist meditation, and Andrew Newberg also noticed a marked change in terms of metabolic activity in the parietal cortex. This is very interesting. Finally, the nuns, who had their eyes closed during the so-called mystical condition, were reporting a lot of visual mental images. We saw activation also in visual portions in the back of the brain, while they had their eyes closed. It seems that these experiences were rather complex and involved much more than a single spot in the brain.

Based on these findings, in terms of EEG, I would like to add something significant. We saw, at the same time, a very marked alteration in terms of EEG activity: electrical activity. There was a very important reduction of brain activity, in terms of electrical activity. We observed a lot of very slow waves, like theta waves, and also delta waves, in the brain of the Carmelites, during the mystical state, which is very interesting, considering that theta waves have been observed in various EEG studies of Buddhist monks as well. Delta waves normally appear during very deep sleep or in some states of coma, but in the Carmelites, there were a lot of these waves prevalent—yet they were fully awake.

If you put all of these findings together, it is clear that, in order to have access to another reality, a spiritual reality, you need to induce an electrical and chemical shift (in terms of brain functioning), relative to the normal states while we are awake. In other words, you need an alteration, at least a slight alteration, to your state of consciousness, in order to have access to these other dimensions, if you will.

■ *What you are saying, then, is that when a person is having a mystical experience, there will be an electromagnetic variation in the brain?*

Exactly, yes, but these things can happen very rapidly and sponta-neously: sometimes a person will seek this type of experience through prayers, contemplation, or meditation; other times it is possible to experience a spontaneous mystical experience. It seems there is no definable trigger, but yet there is an important change in terms of brain activity at those moments. Following the publication of the book, I received many letters from people all over the world, report-ing such experiences, and these experiences occurred apparently without any specific reason. In terms of neurological activity, how-ever, we know for sure that it was something that changed in the brains of people who had experienced these things.

■ *Well, you know there are many stories about people who have an acci-dent and their lives change completely. Many of them become instantly clairvoyant. Another curiosity, since we are talking about electromagnetic intervention, is people who get struck by lightning and then go on to have a completely new capacity for heightened paranormal gifts, or what I would call spontaneous "third-eye activation."*

Do you believe in clairaudience and clairvoyance?

In a certain portion of the brain we discuss these so-called "psychic phenomena." There have been studies for nearly a century now and the evidence for the phenomena is overwhelming statistically, but since they do not fit in the conventional mainstream view in science, they are rejected by several scientists—especially by neuroscientists.

If you believe that what you call "mind," "consciousness," and "spirit" are simple by-products of electrical and chemical activity in the brain then, of course, you cannot be in communication with any-thing outside of the physical body. You cannot receive any information; you cannot emit information either, so you are totally imprisoned within your body and your brain. However, the evidence goes totally against this, so that is why neuroscientists feel very uneasy about those issues, because they simply do not have any good explanation.

■ *That is why we celebrate you for thinking outside the box of convention! I am quite certain that, in your realm of science, medicine, and academia, it took you a lot of courage to come out with such a controversial approach to this very deep subject.*

Well, no, I am not courageous; it is because I have experienced these things myself since my childhood. It started when I was about eight years old, and I am now forty-five. I have had a series of spiritual and psychic experiences throughout my life, and these experiences have clearly shown me that the mainstream view was totally flawed and that we needed a new meta-paradigm in science, and especially in neuroscience. The evidence is there, based not only on my own personal experience but also from the literature available, if you take a look at it clearly and objectively.

■ *So, what kind of psychic experiences did you have as a child and all through your life?*

I have had all sorts of experiences, like clairvoyance and telepathy. My mother is very psychic and very intuitive, so I have experienced the whole gamut.

■ *It is so remarkable that you have gone through those experiences and have been able to bring that firsthand knowledge and experience into balance with science. I am just delighted to see that science and spirit are now coming closer and closer together, rapidly breaking down the barriers from both sides.*

If you take the cases of great scientists throughout history, most of these scientists, like Newton, Descartes, and Galileo, were all very spiritual people—even Albert Einstein.

■ *Well, sure, because they have got to have the inspiration, and the inspiration is spirit, right?*

Exactly, yes.

■ *I have had a number of guests who have talked about various aspects of experimentation with hallucinogenic medications or illicit drugs and also spirit journeying with shamans and hallucinogens. Could you talk to us about what really happens in the brain, from a technical point of view, when someone uses these hallucinogens and what, in your opinion, is really going on with the visions that come from these journeys—let's call them "trips"?*

Based on all our findings and also the findings by other research teams and other research groups regarding the brain in spirituality, it seems that the brain is acting like a reducing value. This is not a new concept. It was proposed about one hundred years ago, by the father of American psychology, William James, and also by Henri Bergson, the French philosopher ... and after that by Aldo Sacslei, the famous writer.

The idea is that, normally, the brain will function very rapidly and, if we take the case of EEG activity, in terms of electrical activity it produces a lot of beta waves or rapid waves. Since we are focused on the external environment, this means that everything that comes from the inside that is very internal, very deep, all the information processed at the level of the unconscious, and also at the superconscious level, is not perceived immediately. We are not in touch with it, because the focus of the attention is directed externally.

I am coming back to the idea that in order to have access to hidden information or information coming from very deep within—in terms of your core, in terms of spirit—you need a shift in electrical and chemical activity.

If you talk about the brain's electrical activity, it means there will be a slowing down of EEG activity, so you need very slow waves, like a lot of theta waves and even also delta waves—the waves I was talking about earlier. If you have access to this, then you can be more in touch with dimensions that are normally hidden, like what we call "spirit" or the "spirit world."

■ *Do you think that a person taking LSD, through the hallucinations experienced, is actually connecting with other dimensions?*

Well, exactly, if you take the case of these drugs, they will target specifically the chemical messenger called "serotonin." We talked earlier about this messenger, and serotonin is crucially involved in spirituality and spiritual experiences. When you take these drugs, you will have an alteration in terms of perceptual abilities, so there is a lot of distortion that is created and that is why we talk about a hallucinatory component in these experiences, but it is not the only component.

The main component, the most important, is the spiritual, because in those cases, the brain is not acting as a reducing value anymore, in the usual sense. Then it is possible for the one experiencing it to have access to other realms. Very often, the experience that is lived under those conditions will be very important for the individual and can have a very long-term impact on this person. I think this is exactly what happens with shamans, for instance.

■ *Have you ever tested anyone that was under the influence of these chemicals?*

No, because they are not allowed, so this kind of research is not possible. Since the late 1970s, it is very rare that we can have permission to do this kind of work, because after Timothy Leary, the U.S. government decided it was not possible anymore to do these kinds of studies.

■ *Because they were doing them themselves, perhaps?*

Well, yes, it is possible. Actually, I have done some experiences myself with these compounds that some people called "entheogens," instead of psychedelics, because the emphasis is put on the spiritual dimension when you use the term "entheogen," instead of the perceptual distortion that I was talking about.

I tested some of these substances on myself and because I have had very intense spiritual experiences before, without taking these substances, I was able to compare. I realized that the nature of these experiences is very similar. This has been studied in the 1960s and the beginning of the 1970s, by some people; there is a great theological historian, Huston Smith, who wrote several books on perennial philosophy and the commonality between all the great spiritual traditions of the world.

In 1962, he was a theology student at Harvard. He participated in a famous experiment called the "Good Friday Experiment." The lead scientist was Walter Pahnke, who was working with the effects of psilocybin mushrooms. They did a comparison between a placebo group and an experimental group. Huston Smith was a member of the experimental group and at that time, during this experiment, he had a very intense spiritual experience. He is now more than eighty years old. The effects of this specific experience lasted over fifty years.

So you see it is not true that, because you use a chemical means to induce an alteration in the state of consciousness, that automatically the subjective experience is any less genuine.

■ *Did the people that were on the placebo still have the effect?*

No, they didn't have it at all. For the placebo, this effect is not very long term; after a few months usually it will disappear. People need to know that when you induce an experience—for instance, using other means like prayer or contemplation or meditation—you automatically induce changes in terms of brain activity.

■ *Can you elaborate that, because I think that people really need to hear that.*

Well, you can do all kinds of things. Anything, for instance, if you breathe on a very regular schedule, for instance like what they do in rebirthing ... or if you pray on a regular basis, you change your brain. There are neuroscience studies coming out now showing that if you meditate on a regular basis you will change even the structure of your brain. So, of course, you change everything when you do these activities: you change what is going on electrically, chemically, also in terms of your own brain structure.

■ *It's like a workout for the brain. Brain gym!*

Exactly, yes.

■ *All right, now here is a question for you. We hear more and more that people who are having transplants of organs, the receiving people, start to behave very strangely, acquiring personality traits of the people they have received the organ from. It was recently in the news: a woman, I believe she committed suicide after she got the heart of someone who had committed suicide. She got the heart, then she committed suicide. There are more and more stories about this, that are documented, about personality traits, eating habits, etc.*

I would like you, from your point of view, to describe how you feel that this is being communicated through the mental body. Let's say that it is a heart, because I believe that the mind is in every cell of every organ of every body—in all the tissue.

Well, yes, that is the most likely explanation. I think that when we refer to "mind," for instance, we should see it as an informational shield. It is possible that the memories of the donor have affected all the organs of her body, before she committed suicide.

■ *Sure, because by nature of the creation of the physical body, every cell has a command that it responds to; it has a knowledge of what it is supposed to do.*

Exactly, so we have done a series of brain imaging studies in our labs, showing that it is possible, simply with a very specific intention, to change the brain activity chemically or electrically—simply by intention and volition and desires. The neurons—the nerve cells in the brain—will respond very specifically to these intentions, desires, and emotions. It is probably exactly the same kind of phenomenon.

■ *This to the layperson would be recognized through the term "mind over matter."*

Yes, exactly.

■ *There is so much stress and turmoil at the moment and a lot of people are getting into fear and lack, the lower vibrations of thought and emotion. I feel it is important, through this conversation with you, that not only from a spiritual point of view, but also from a scientific point of view, we always remember how our thoughts directly affect how we manifest our reality.*

It might sound hackneyed, but I think it is a message that we have to keep hearing and hearing and hearing again. So you are saying that, with the process of intention and meditation and volition and will, we can actually change the entire structure of the brain for a higher purpose?

Yes, and it is **even** possible that, if you do this, you will even alter your gene expression for specific genes. It has not been shown yet, but we expect that this might be the case. I am sure that, in the next few years, we will be able to demonstrate this. At the same time that you demonstrate what is going on in your body physically, you also influence what is going on externally to you, because there is increasing evidence in side research showing that humans have the capacity mentally to influence electronic systems, biological systems, plants, animals, and other human beings as well.

It is like we are all interconnected within a sort of unified field and this field is part or comprises everything in the Cosmos, in the Universe.

■ *We are relearning this because we have been through such a period of separation; an age of logic, reality—let's call it the "Age of Reason," even though reason rarely seems to prevail! There is a lot of separation in that approach to our existence and we are relearning how everything is connected. Do you agree?*

Yes, totally.

■ *We have to learn really quickly here! What do you feel is one of the most important messages that emerges from your work?*

I started as an independent researcher in neuroscience about twelve years ago and since then I have done a series of brain imaging studies. If you take these studies all together, they all demonstrate that human beings are not biological robots, totally determined by their neurons and by their genes. To the contrary, like you said before, we have to be very careful in terms of what we think and our own feelings and emotions, because they have the power to totally change your brain activity and the functioning of your brain, in the physiological systems that are connected with the brain. They can also influence other people and everything that lives on the planet. In other words, humans have an enormous power, but they are not aware of this yet collectively. So, that would be the take-home message.

▪ *I agree with you completely, and through my work this is the most important message. I do my best to help people understand how they create their own realities and that reality is a reflection of what you put out into the environment—immediate and beyond. I resonate with the Buddhist path, which teaches you to take responsibility and credit for what you create. That helps us to truly understand how you are using thought.*

Yes, because thoughts are energy, so we have to be aware all the time. We are much greater than what we think we are.

▪ *Indeed. We need your research and more research to help the skeptics understand, by being able to see how this works, in what they call "black and white." We need to help the skeptics get past the doubt and into the light, the stream of light that we are working to bring forward for humankind. Surely your studies, where you are actually able to record the patterns of a spiritual experience in your tests, as contrasted with normal brain processes, will allow people to recognize the differences, the alterations of the brain in the spiritual or "mystical" experience.*

I am sure it will come during the next decade.

■ *On another subject: I think we are going to have a revolutionary moment very soon now. I do believe we are going to have contact with extraterrestrial intelligence. I think we have had it already; I think we are going to see something happening around the globe that is going to be a very important trigger for humankind.*

Well, you already have some evidence of this with the crop circles all over the world.

■ *Yes, the crop circles; have you been in one yet, Doctor?*

No, never, but we have to remember that humankind is still very young from an evolutionary point of view. It is like if we are in kindergarten, but there are many other beings in the Universe in various dimensions trying to help. I am saying that based on my own experiences.

■ *I am coming to Montreal in October next year. May I have the pleasure of meeting you?*

Absolutely.

■ *I would love to. I know it would be a wonderful encounter. I am just so enthralled with all that you bring to the table, as far as your personal experience and your incredible academic and scientific knowledge. Once again, your book is called* The Spiritual Brain: A Neuroscientist's Case for the Existence of the Soul. *It is a fabulous book.*

The back cover asks the question: "Did God create the brain, or did the brain create God?"

Do you have an answer to that, Doctor Mario?

Well, it depends on the point of view. Humans create their own concepts about God and have always been doing that for thousands of years—so you see some differences in terms of the great spiritual traditions of the world. But at the same time, I believe that everything that has been created—everything that is "organized" physically—needs an intelligent designer.

There is more and more evidence coming, for instance, from cosmology and physics, showing that even the slightest alteration (in terms of the constant of physics) means you don't end up with Planet Earth and a physical universe like the kind we have. So, it is clear and very obvious that you need an intelligent source at the origin of everything in the Universe.

My response is "yes," you can use the term "God" if you will, but yes, God has created everything that is physical, including the brain.

■ *God ... the Great Mathematician ... the Supreme Architect: the Prime Creator. Whatever term we use, what I hold close to my heart is the knowing that there is a most exquisite order to what appears to be random and chaotic in the Cosmos of Soul.*

At this time of our Great Shift as individuals, as societies—as a species— it is so empowering to remember that we are co-creators of that divine order and that, despite the illusions of our struggling planet ... all is progressing according to some undefinable, but exquisite divine plan. All is perfect, according to that Great Intelligence—Creation.

■ ■ ■ ■ ■

The book is: *The Spiritual Brain: A Neuroscientist's Case for the Existence of the Soul* by Mario Beauregard and Denyse O'Leary.

The Giza
Power Plant

■ ■ ■ ■ ■

Christopher Dunn

Author of the best-selling book on the pyramids, *The Giza Power Plant: Technologies of Ancient Egypt,* Christopher Dunn is an engineer with over forty-five years' experience. His pyramid odyssey began in 1977, when he read Peter Tompkins' book *Secrets of the Great Pyramid.* His immediate reaction to the Giza Pyramid's schematics was that this edifice was a gigantic machine. The purpose of this machine involved a process of reverse engineering that has taken him over twenty-four years of research to discover. One of the important clues of his theories about the Great Pyramid has to do with the precision of ancient Egyptian granite artifacts, which Chris believes provides the smoking-gun evidence of a civilization existing in prehistory—a civilization far more advanced than previously believed.

■ ■ ■ ■ ■

Many of the bright thinkers of our time are looking back to the wisdom of the ancient engineers, designers, architects, and scientists in a quest to understand how some of the great works of antiquity were created, for what purpose, and with what eye to the future of humanity. Certainly the Great Pyramid of Giza, the only one of the Seven Ancient Wonders of the World remaining, still manages to defy history and the countless generations of seekers who, throughout time, have searched to find the answers to these questions—still taunting humanity, still shrouded in mystery.

Where do we begin, Christopher, to explore your years of profound study and investigations into the enigma of the Great Pyramid and what it has meant to humanity—perhaps the planet itself—throughout time?

When I looked at the huge question mark surrounding the Great Pyramid, the initial inspiration for a possible solution as to its purpose was not difficult. But then, gathering all the evidence to support the original theory took many, many years. It was basically more detective work than sheer genius, and a lot of synchronicity along the way, with people coming in and out of my life, bringing in different bits of information. The book *The Giza Power Plant* has my name as the author, certainly, but it wouldn't exist without a lot of other wonderful people helping.

And we thank you and all those who have contributed to your discoveries, because it really is a significant work, I think, for our generation. I really feel strongly about it. We will get into the revelations in the book in a minute, but I would like to start by celebrating the idea that you dared to come out so boldly against the explanations of conventional Egyptologists.

Well, it's not my intention to persuade the community of classical Egyptologists, but rather to actually provide some new ideas or a different way of looking at things for people like me, who are just laypeople, or people who have some experience in technology. I hoped that maybe engineers or technicians, or even people without any background in engineering, could actually work through the book, understand it, and see at least if they could see the threads of evidence that led to my conclusions about the purpose of the Great Pyramid—even if they do not agree with the book.

■ *Don't you find, though, that having that "different way of looking at things" in Giza can make it more difficult and tricky for you to get doors open—so that you can conduct your research?*

That's a very good question. It can be a little "tricky," as you say, and that is probably a good word, but actually the way I approach it is pretty *laissez-faire*. When I go to Egypt, I take some instruments, with the intention of accomplishing certain research. I find that some doors are closed to me and even though I may have a prior arrangement to do something . . . it is not possible.

For instance, I find it very, very hard to get into the Serapeum at Saqqara. I found myself in there in 1995, when I was on location with Netherlands Television with Graham Hancock and Robert Bauval, for a documentary we were doing. They had permission to go in and I went along with them . . . and what I found there was unbelievable to me.

For me, a manufacturer and somebody who has worked with precision machine tools for many, many years, it was startling to find evidence of advanced machinery and technology, used in the production of huge granite boxes. Since the manufacturing capabilities of the ancient Egyptians have been a consuming interest for me, being in the right place with the right person at the right time was very significant.

On my Web site, I show photographs of me inside one of these granite boxes, with a precision toolmaker's square and straightedge, actually doing a comparative check on the flatness and squareness of the box.

These boxes were cut out of a solid piece of granite and the surfaces on the inside, that I was checking, had the same precision that you would find on a modern manufactured surface plate. They were very precise. The gauge that I used had a precision of one-twentieth of the thickness of a human hair. That is extremely precise and more … sliding it along the surface of the granite surface on these boxes, I found that there was no variation in the surface at all. It was perfectly linear. Even shining a light, looking to see if there were any gaps (even one-one-thousandth of an inch would show), I didn't see any variation whatsoever.

■ *That is about as precise as it can get, I would suppose! And, of course, it definitely defies the standard explanations of how the ancient Egyptians quarried the stone.*

Indeed. The other amazing thing is that it is not just the flatness but the perfect squareness that illustrate their sophisticated manufacturing capability. In the photographs, I show the square on one surface of the box—the underside of the lid—and then directly opposite on the inside wall, where you see the same condition. This indicates that those two surfaces—those two walls on the inside of the box—were crafted perfectly parallel to each other, because otherwise that condition would not exist.

That is an unbelievable feat even by today's standards! I went to several granite manufacturers, experts in precision granite manufacturing. Only one actually came back to me and said that they could not reproduce this box; they simply don't have the equipment to do it.

■ *One of the things I find very exciting about your work is that you con-*
stantly refer to the fact that not even modern technology can, in many
cases, reproduce what the Egyptians were doing, who knows how many
thousands of years ago.

I would not state emphatically that we **couldn't** do it. It is just like
the Hubble Space Telescope; it did take an extraordinary effort. It
was conceived of and it was accomplished. In the process, new and
different tools were designed and built, just to accomplish the Hubble Space Telescope. So, if we had a purpose or a reason to do this,
we could probably design the tools and the means and the methods
to actually produce it, but the investment would be enormous.

Just the raw block of stone would cost over two hundred and fifty
thousand dollars and to machine it out and finish it out on the inside,
to this order of precision, would cost an unbelievable amount of
money. I can't even fathom how much it would cost to do that.

■ *Plus we have not only this granite box that you are talking about, but*
so many of the "coffins," let's call them, or the sarcophagi, which are also
granite carved.

Even in the tunnels at the Serapeum, where this box is located, there
are over twenty others. So, it's not just one granite box. They were
accomplishing this kind of work in many different boxes: in the King's
Chamber in the Great Pyramid and, particularly, in Khafre's Pyramid there is a granite box in the so-called "King's Chamber" that
has an equal to what you see here.

Unquestionably, the ancient Egyptians were masters at working
with this granite.

■ *So how do you feel they accomplished it? What does your intuition or*
your research tell you?

My intuition is actually based on my experience in manufacturing and looking at the features, as there are certain features on these boxes that indicate the means of manufacture. One of the more telling ones is the corner radius where the surfaces come together and the corner is very, very sharp—extraordinarily sharp for something this large—and you have to ask the question: "Why did they have to have it that sharp?'

To accomplish this, I believe, they had to have had some kind of guided tool that was very precise and that actually cut the surfaces on the inside, the same way that we would have a machine tool that is guided along a granite base ... because we actually use granite today for some machine tools. This is particularly true in the case of inspection equipment, like the coordinate measuring machine,* because granite is a very stable material. It does not shrink or warp, and there is no variation in the surface, once you get it finished to a perfect flatness.

■ *What about the possibility that they were using some form of laser to do this cutting?*

The laser is not really an efficient tool for cutting stone, because laser is a thermal process and not effective for cutting stone—particularly granite—since it is actually an igneous rock, and so would withstand a lot of heat. I started in laser manufacturing in 1983, and we were working with high-powered YAG lasers and CO_2 lasers and we were cutting aircraft-type materials, exotic metals, but not stone.

■ *Well, what do you think they were using?*

*A metrology machine that navigates Cartesian coordinates (x, y, and z) to gather dimensional data from a manufactured item and verify its conformity to certain specifications.

I think they were using special grinding equipment. Essentially, what they would have used would have been a diamond, probably a rotating disk that was grinding the surface and actually moving along an axis and finishing it with precision. There is evidence, too, in other areas (not just in these granite boxes, but also on the Giza Plateau). We find, out in the field, evidence that shows that objects would have used a multiaxis machine to create the contours on granite cornices.* There is a beautiful granite cornice on the south side of the Valley Temple.

You've been to the Valley Temple, I am sure. There, on the south side of the Valley Temple, is a granite cornice just lying there, which was originally on top of the wall. This is an object that has a three-dimensional profile. The profile consists of two radii at a tangent with each other—they flow together—and they are extremely precise geometrically. How did they create that?

Egyptologists will tell you that they used dolerite** pounders to pound the rock.

■ *I just saw one of these historical enactments presented on the Discovery Channel, where they depict how the Egyptians created points of impact on the stone and then, once the fracture lines had been determined, they would pound it and it would crack in half. It's an interesting theory, but it certainly doesn't explain one-tenth of the thickness of a hair precision!*

No, and they never do, but think, since they can show one stone and beat another stone and then remove some material, that they've solved the problem ... but they haven't solved the problem, because they haven't really explained anything. A child could do that.

---

*Curved architectural elements normally placed atop Egyptian temple walls or doorways.
**A greenish stone that is harder than granite.

■ *Why would it be so hard for the conventional thinkers to embrace the idea that a higher form of technology was known to the Egyptians, at a certain point (or points) in the evolution of that civilization? You would think that the classicists would be proud of such accomplishments, no?*

I think that is an excellent question and I wonder, too! I particularly wonder why the modern-day Egyptians don't embrace the idea that their ancestors were highly advanced technologically. It is amazing to me that they don't look at this, but that they still actually accept the Western Greco-Roman historical model, which puts the Egyptians further down on the evolutionary scale. It would indicate that the Egyptians were not as advanced as the Greeks or the Romans were, but the evidence on the ground in Egypt shows us something entirely different, in my opinion.

■ *I think it might also have to do with the fact that the classic Egyptologists just don't want to even consider that there might also have been an influence from another highly evolved civilization, perhaps Atlantis, that could have come in and given a jumpstart to the civilization.*

In one sense, I think that is where the mistake has been made (and basically what a lot of people have said is that the Egyptians of four and a half thousand years ago were not capable of this kind of work). I mean, obviously, if you look in the ancient Egyptians' toolbox, you don't find the tools that were capable of doing this kind of work, and so it is easy to jump to the solution of a different culture coming in and helping them out, whereas there could be another answer to that question.

■ *It could also be that the "tools" have been deliberately removed, so that the officials can hold on to their timeworn explanations of the Egyptian civilization.*

Yes, "removed," perhaps, or actually over time destroyed or carried off—for all those reasons. What matters is not just that the tools have been destroyed or buried, but the knowledge of how they worked has been lost.

I definitely believe, in my heart, that there was a cataclysm in the past that destroyed large segments of civilizations on the planet, and that those who survived went through enormous suffering . . . essentially unable to ever rebuild their society back to what it had been. Ultimately, generation after generation, they lost the knowledge, and sank into a primitive or very simple agrarian-type society.

■ *Let's not forget, if we look at our contemporary society . . . let's talk about one hundred to a maximum of one hundred fifty years ago, we didn't even have electricity. We observe and study Egyptian civilization, spanning thousands and thousands of years, in comparison to what enormous feats we have accomplished in our modern society, and look at how much potential time there was for the sophisticated societies to come and go, in the history or timeline of the many civilizations of Egypt.*

Absolutely, and I think that is where the arrogance of the conventional Egyptologists may stem from—that belief that Egypt was the first and the only advanced civilization that existed on the planet. I think there may be another reason for that. It may not just be chauvinism; it could be real fear. Imagine the implications of actually knowing or understanding that civilizations are actually mortal and can be destroyed! Nature is cyclical. Our current civilization may very well be facing this possibility very soon.

■ *Well, hopefully later, rather than sooner . . .*

Hopefully later, yes. You know, with the way the Universe works, it seems that civilizations over the centuries have constantly risen and fallen and, of course, we do want to escape a global catastrophe. I certainly find it a comforting thought to think that we could escape it, and I certainly would contribute to any effort to escape it. I don't want to die in a ball of fire or in a tsunami! But when you look at what has happened in the past, perhaps from the clues of ancients, I do think that we have to take measure of our lives and say: "Okay, how do I want to live my life?"

This could happen. Perhaps this is one of the greatest lessons we can learn from the past.

Yes, I agree. I really feel it is important for people to understand that every day is worth celebrating. Any day could be your last day, and you would want it to be as filled as possible with light and love and celebration, rather than the dread and doomsday mindset that seems to be dominating the consciousness of now.

Shifting gears now, I would like to ask your insights into the Coral Castle mystery. For those who maybe don't know about Coral Castle, it is a megalithic structure built in the early twentieth century by a man by the name of Edward Leedskalnin, in Florida. He created a huge structure, with impossibly placed megalithic stones, and nobody knows how he did it.

He used to work at night, alone. People would hear strange sounds and all kinds of things going on, but he never revealed how he did it. Unfortunately, he passed away and took his secret with him.

There's no question he utilized some form of hidden knowledge to lift the stones—he didn't have the technology to accomplish it by any conventional means, to be sure! I have my own opinion as to how he did this, like the ancient Egyptians and others who lifted these giant stones. I think it involves sound—reaching the appropriate musical frequencies of the stone—but I would like to hear your opinion as to how one man managed to create a modern megalithic site on his own.

Well, I do believe that Edward Leedskalnin definitely knew the secrets of how the pyramids were built. He said those very words himself. He is the only person that I have read about or heard about who actually has done something that you could look at and say: "This fellow was definitely on to something."

He is reputed to have weighed less than a hundred pounds! He was a small fellow from Latvia and over a course of about twenty-seven years, he built this castle in Homestead, Florida, out of monolithic coral blocks of stone. The heaviest of the stones he lifted was a thirty-ton block, but there were other items that were huge as well: an obelisk that weighed twenty-three tons, a nine-ton gate, and other extremely large objects.

Coral Castle really is an odd place—well worth visiting. I definitely recommend that people visit it. It's a really fascinating place.

■ *What do you think, Christopher? How did he do it?*

My feeling is that he used a combination of sound, vibration, and electromagnetism. I think that he had some kind of machine that enabled him to affect the natural magnetism of the Earth, in some way reversing the polarity and reducing the gravitational pull on the coral blocks.

Now, why do I say that, knowing it is kind of crazy? Well, we are freewheeling and out of the box in this conversation—and this fellow was freewheeling and out of the box, too. He had a different view of nature than the scientists. He wrote a couple of pamphlets (one on electricity and another on magnetism) based on his understanding, or what he believed, about nature and electricity. He believed that gravity was just individual magnets that were being emitted from the Earth. Let's be clear about the nature of magnets. They have a north and a south pole, and the magnetic properties are described as "like" poles, which do not attract each other, and "unlike" poles, which do attract. Therefore, north and south would be attracted together—but he believed that these individual magnets moved through space and matter.

If you get inside Leedskalnin's head, you think: "Well, if that is true, then did he find a way to reverse the polarity of the individual magnets in the coral so that they were aligned and in opposition to the individual magnets emitted from the Earth?"

■ *In other words, matter that was in the field of the reversed polarity would virtually become weightless?*

Yes. It would become weightless and float. It is speculation, of course. Some of the photographs that you see of him at Coral Castle show him working with several very large tripods. These tripods have a box on the top of them and wires running down from the top and on one of the tripods is a chain hoist for lifting.

Looking at the chain hoist, the chain that I measured was rated for about ten tons—so it would have been incapable actually of lifting the thirty tons or the twenty-three tons that he had shown he had lifted. So, it is a real enigma … a real puzzle! People speculate on how he might have done it and some have actually tried to replicate some of the equipment that he has, but, to my knowledge, nobody yet has actually been able to duplicate his feat.

■ *One wonders why didn't he leave us this wisdom? Why did it have to disappear with him?*

My sincere feeling on why he took the secret to the grave is because he was a product of World War II: he didn't trust governments; he didn't trust people. He was afraid that his secret would get into the hands of the military and the government.

■ *That would make sense ... that would be a fear that would probably be warranted. Imagine anti-gravity in the hands of the wrong people.*

True, but can you imagine a world **with** that kind of technology? Take the fuel situation, for example. We've got gas going over four dollars a gallon ... but if we had this technology, we wouldn't need that much fuel anymore. Essentially the reason why we burn so much gas and use so much energy is to overcome gravity. That's it in a nutshell. When we get in our vehicles and we drive down the road, we are propelling and pushing against the forces of gravity. So, if you can overcome the gravity, then you don't need as much energy to move around.

■ *Imagine the responsibility of such knowledge....*

Yes, he was tapped into something. He was a very odd fellow, and I have no idea as to what was going on in his mind, but he wrote a little book called *A Book in Every Home*, which is a real oddity for those that have read it. Supposedly, it is an instruction manual on how to raise a young lady—in quintessential male chauvinist terms, I might add. It is an odd book and I just can't help but think there is some kind of code or something hidden there. Did he leave his secret, encoded? We just have got to figure it out.

■ *The obvious question is: do you believe this is how the huge stones were lifted for the pyramids?*

Most people have some kind of an opinion on how the stones were lifted in Egypt. I think that whatever technique he had developed there held the clues to the abilities of the ancients. I believe we can say with confidence that he was using some form of technology that the ancients used to build the Great Pyramid and some of the other structures in Egypt: moving monolithic pieces of stone weighing five hundred to a thousand tons and lifting all the granite beams above the King's Chamber, seventy-five feet in the air!

■ *It does exhaust me to continually see these documentaries depicting really primitive hauling of these stones being dragged into position by hundreds of slaves.... I wonder how long will this myth persist, and why anyone would believe it.*

It's true. There have been quite a few documentaries that purport to tell us how the ancient Egyptians built their pyramids and moved stones up there. In those documentaries, you always see a bunch of guys in diapers pulling on a sled with a half a million ton block on it, which is really funny. Normally I turn those documentaries off. I don't even like to watch them because they are just so unrealistic.

■ *I saw a similar one recently about the blue stones at Stonehenge. The same thing—these stones were supposedly dragged along I don't know how many miles of land with these funny little man-made rolling machines, out of wood. That's a seven hundred ton stone on a slatted wood pulley!! Please! It's time for some more substantial information to go mainstream.*

I agree. I think people around the world would really like to know the answer to how these feats were accomplished, and that they are not satisfied with what they are told and what they see on these "official" documentaries. A wealth of information is available to point to and say: "If you are going to create a documentary to show how the ancient Egyptians built the pyramids, take the most difficult aspects of the work that you know of and replicate those."

They never do. They take something simple and small for proof, and jump to conclusions: "Okay, we have been able to move this two-ton block with five people, so we can move a one thousand ton block if we just add more people!"

▪ *Right ... but seven thousand people later they've still got only one stone lifted.*

Chris, we've got to jump now to the big question. In your opinion, given that neither you nor I believe the story that the pyramids were tombs—what were they? What was their purpose?

Well, the title of my book is *The Giza Power Plant* and it clearly proposes that they were used to generate energy. We have to consider what this power was ... what this energy was. In the book, it is described as "natural energy" that is drawn from the Earth and converted into microwave energy, through a process in the King's Chamber. The book actually uses all the evidence that we find inside the Great Pyramid to support its central theory.

Since writing it over ten years ago, more insight and information has come to me, so I am very encouraged. But at the same time, I think that the true benefits from this kind of a power system are not necessarily just in the production of energy.

What I have come to believe and understand is that the greater function of the power plant was to harmonize the planet. In other words, it was made to drive the tectonic plates into resonance and to relieve seismic stress in the Earth, which, at the same time, created harmonious frequencies for human beings, relieving stress in human beings as well. So you could have a totally holistic power system that energized the population on many levels.

The pyramids were actually acoustic devices that could generate certain vibrations within the Earth, and these vibrations created harmonious sound waves within the Earth and aboveground.

■ *In fact, I have brought a lot of groups into the chamber and we have done a lot of truly remarkable harmonic toning, and what happens is just so remarkable. The acoustics in there are utterly unbelievable. We do collectively even see beings appear there on the waves, so I would suggest that these sounds, reverberating within the structure, somehow bring the dimensions into harmony as well.*

I too have been in the Great Pyramid many times. If you haven't actually been inside the Great Pyramid and experienced it, then it is very difficult to understand it, but there certainly is a profound effect on the human organism. It can actually really stimulate creativity. It can affect you on many different levels, where you really enjoy a feeling of cosmic consciousness and Oneness with your fellow man and the Universe. There is no place like the Great Pyramid to achieve that.

■ *. . . Except perhaps in outer space, as it happened to Edgar Mitchell!*

Perhaps!

■ *I've got a little story to tell. You said you got sick in Egypt, and so did I. Many years ago I developed pneumonia when I was there, but still managed to climb into the Great Pyramid with Franco, my partner, as we had booked the entire pyramid for private entry. I couldn't let the opportunity go, so I braved it, trying to suppress the incessant coughing, so that he could have an experience in the sarcophagus. I had been inside so many times, but he had never been there and the noise of the intense coughing reverberated all through the pyramid! To say the least, it was a distraction for us both.*

When we got back to Rome, I had to go immediately to the hospital. When the doctor showed me the X-rays of my lungs, he said: "As you see here, there is a pyramidal form of pneumonia in your left lung." He didn't know that I had just returned from Egypt.

I said, "Excuse me, how did you describe that?"

He answered, "There is a pyramidal structure in your lung."

I thought, "Wow. I bet the pyramids somehow contained that pneumonia from spreading and that that geometry—one of containment—was reproduced in the consciousness of the bacteria or virus that had created the pneumonia in my lung." When he showed me the X-ray of my lung, which I hope I still have, it was, in fact, the pyramid—perfectly geometrical. What do you think of that?

Well, I believe you, I really believe you. At one time in my life I wouldn't have, but I have now experienced enough and seen enough and felt enough in Egypt that I definitely do now.

▨ *In fact, on my near deathbed I just listened to him, I had no strength left at all, but when he said there was "this pyramidal structure" in my lung, it definitely got me sitting upright and paying attention.*

No doubt! That is fascinating.

▨ *I just had to share that with you, as it ties in with your comments regarding the vibrations emitted by the pyramid. Can you elaborate more for us about your theories of the Great Pyramid as a power plant: a real synopsis of your thesis, your theory, and how this worked?*

Essentially the main features of the pyramid are of course the mass of stone covering thirteen acres and the internal features:

- the Subterranean Chamber
- the descending passage that goes down to the Subterranean Chamber
- an ascending passage
- the Grand Gallery

- the King's Chamber
- the Queen's Chamber, which lies below the King's Chamber

The conventional theory is that the Pharaoh Khufu commissioned the pyramid to be built and so they built the Subterranean Chamber. He changed his mind ... then they built the so-called "Queen's Chamber." He changed his mind again ... and then they finally built the King's Chamber.

My theory takes into account all the features within the pyramid, especially the precision, because if you look at the cross section of the Great Pyramid, the schematic looks more like a machine than it does any other kind of tomb or structure. I was intrigued that the passageways are so small (about forty-one inches square) and that the chambers have shafts running off them at an angle. And there are certain anomalies inside the King's Chamber and about the King's Chamber. Then, of course, the state-of-the-art manufacturing and the building of it ... everything is so precise. Just consider: the descending passage, which goes one hundred fifty feet into the ground, below the pyramid, is perfectly straight within two-hundredths of an inch—or the thickness of a thumbnail.

These amazing features are what actually set me on my quest to try and understand further how the pyramid worked.

My thoughts were that it had to be some sort of a power plant. It was the largest structure on the planet for thousands of years and there had to be another reason why someone would build something so huge, besides the story that it was to bury a king. That just didn't make sense to me.

■ *It doesn't make sense to me either. Let me get that statistic again?*

The descending passage, over one hundred feet through the constructed portion, was measured by Petrie in 1883 and found to be within two-hundredths of an inch—the thickness of a thumbnail—of being absolutely straight.

◾ *That is remarkable, Chris. These are the kinds of facts that can only make people sit up and take notice of alternative theories, such as yours, as to the true purpose of this immense structure.*

It **is** remarkable! Throughout the inside of the pyramid, such as the limestone blocks in the Queen's Chamber, the stones are fitted so perfectly that even a piece of paper would not fit between them. It is beautiful stone, very finely finished. And so, essentially taking all the features, all the evidence, within the Great Pyramid, I renamed the features.

The Subterranean Chamber is still the Subterranean Chamber. The Queen's Chamber became the "Reaction Chamber," because of the salt substances that were found on the walls. I propose that these shafts leading down into, but not connecting to, the chamber delivered chemicals and that there, in the Reaction Chamber, the chemicals mixed and created hydrogen. The hydrogen flowed upward into the King's Chamber, along with sound waves that were generated by resonators in the Grand Gallery, which I renamed the "Resonate Hall." This explains the twenty-seven pairs of strange slots that are located along the length of the Grand Gallery. The resonators in the Grand Gallery were affected by vibration flowing through the mass of stone, converting the vibration into airborne sound.

That sound was focused into the King's Chamber, which is constructed of pure granite. I call the King's Chamber the "Resonant Quartzite Chamber," because of its acoustic characteristics—it being built free and independent from the rest of the structure. It is actually not tied into the core masonry and it vibrates freely and resonates freely. This has actually been tested by acoustic engineers, who have determined that it actually has resonant frequencies that form an F-sharp chord. These are but a few of the clues given to us by the Great Pyramid which support the power plant theory.

Some people would say that perhaps they ancient Egyptians weren't using it like a power plant as we would consider it today—in terms of creating electricity to fuel our civilization, operate our toasters and microwave ovens, and that, perhaps, it was a subtler energy … maybe a communication device.

■ *Or maybe the ancient Egyptians were harvesting the energies of the planet.*

The core basis of the theory is that they were harvesting the energies of the planet, driving the planet to "resonance," to vibrate at a particular frequency.

■ *That is exactly in tune with what I bring forward in my work of studying and teaching the aspects of sound and its effect on the microcosm as well as the macrocosm—from cell to space—so I am particularly pleased to hear you make that statement.*

What about the Subterranean Chamber—it is an unfinished space. How do you describe that?

The Subterranean Chamber is an incredible, carefully designed chamber. One might think that it wasn't finished, but indeed it was. It doesn't look like it is finished, because it is rough and it doesn't have any particular design, or recognizable design, to it—in other words, it's not square, it's not round. It's got deep gouges in the floor; it's got these unusual "baffles" (raised half walls) toward the west, but the chamber is an essential part of the machine.

There is a gentleman named John Cadman in Bellingham, Washington, who was as consumed about the question of the Subterranean

Chamber as Richard Dreyfuss was obsessed with the mountain in *Close Encounters,* when he was building his mashed potato model.

John Cadman was getting up at three o'clock in the morning and sitting in his living room, carving models of the Subterranean Chamber out of plastic. He spent years actually doing research on this and what he actually ended up with, in his words, was a "pulse generator." The Subterranean Chamber, he declared, is a pulse generator. He sent me all his drawings and his calculations of how it worked and I thought: "Well, that's really nice, but I need to actually witness it and see it for myself."

I had to go to visit him, as I didn't want to dismiss it offhandedly without actually witnessing what he was talking about, so I flew out to Washington specifically to find out. I did it while a hydraulics engineer, Dr. Jack Cole, was in the area; he was there just to witness it too.

Patricia, it was amazing. What he had actually done was to reverse-engineer the Subterranean Chamber. He had the descending passage down into the chamber and water flowed in. He turned on a valve, like a faucet, and the water actually started to flow up a pipe. He choked off that valve and opened up another one and it flowed down to a waste gate. As soon as the waste gate, which was actually a little flapper valve, started operating, the ground started to thump! This thing was going thump thump, thump thump, thump thump. It was beating like a heart.

■ *Was he picking up the heartbeat of the planet?*

You know, I can't help but make that connection . . . that you could tune the Great Pyramid to the heartbeat of the planet. The thing was a very holistic power plant. They were feeding a heartbeat into the Earth's plate to gently release the stresses that build up in the Earth's crust (where the plates meet).

■ *Do you think we can reactivate the pyramid, today, to perform that majestic feat of calming the Earth's imbalances?*

I believe it would be worthwhile to draw the necessary talent to work on building these technologies and applying them today. I doubt that the Egyptian government would ever make the Great Pyramid available for this kind of restoration. No doubt, the Great Pyramid has a lot to teach us!

■ *What do you think about the desire of the government to cap the pyramid, back at the millennium?*

I thought it was much ado about nothing. Common sense finally prevailed and a stop was put to it. I don't think it was because of potential damage to the pyramid. The political forces in the country saw it as a sort of media madness with all manner of egos pushing their own agendas.

■ *Do you have one final thought that emerges from your work and discoveries?*

Yes: respect the builders of times past. They were not as unsophisticated as conventional documentaries would have us believe.

■ *And we, Chris, are dedicated to helping expose the wisdom and the wonder that lies beyond the box of conventional thought, working to shine the light on all that has passed before us, all that is great in our world, and all that lies ahead....*

■ ■ ■ ■ ■

To find out more about Christopher Dunn, author of *The Giza Power Plant: Technologies of Ancient Egypt,* visit his Web site at: *www.giza-*

power.com. Available July 2010: *Lost Technologies of Ancient Egypt*, in which Christopher continues his quest for more evidence that will elevate the ancient builders among all nations and epochs of history. Karnak, Dendera, Luxor, and Abu Rawash hold breathtaking new evidence of a highly advanced, ancient culture.

The Spiritual Technology
of Ancient Egypt

■ ■ ■ ■ ■

Edward Malkowski

Since the dawn of the Age of Science humankind has been engaged in a methodical quest to understand the Cosmos. With the development of quantum mechanics, the notion that everything is solid matter is being replaced with the idea that information or "thought" may be the true source of physical reality. Such scientific inquiry has led to a growing interest in the brain's unique and mysterious ability to create perception, possibly through quantum interactions. Consciousness is now being considered as much a fundamental part of reality as the three dimensions we are so familiar with. Although this direction in scientific thought is seen as a new approach, the Secret Wisdom of the ancients presented just such a view thousands of years ago. Building on René A. Schwaller de Lubicz's systematic study of Luxor Temple during the 1940s and '50s, Edward Malkowski shows that the ancient Egyptians' worldview was not based on superstition or the invention of myth but was the result of direct observation using critical faculties attuned to the quantum manifestation of the Universe. This understanding of reality as a product of human consciousness provided the inspiration for the sacred science of the ancients—precisely the philosophy modern science is embracing today. In the philosophical tradition of Schwaller de Lubicz, *The Spiritual Technology of Ancient Egypt* investigates the technical and religious legacy of ancient Egypt to reveal its congruence with today's "New Science."

■ ■ ■ ■ ■

■ *Ed, how did you get involved in such a monumental task of writing such truly scholarly works as you have?*

Well, it's a search for truth! It started at an early age, probably about the age of eight, when I became aware of my own eternal life and started asking the questions everyone else asks: Who am I? Where do I come from? Why am I me and not somebody else?

■ *I'm not sure everybody else is asking those questions at eight, Ed! . . .*

That may be true. For some reason, I have this thirst inside of me to know why we are here and what we are doing. Around that time, I became interested in history. My dad started bringing home magazines about Marshall Cabbages' *History of World War II*, and I became interested in that and it stayed with me. In junior high, I delved into U.S. history and, in college, more into U.S. history and then moved into ancient history. It just seems I have grown up with this thirst to know how civilization started. As for the scientific part of it, I'm not really a scientist, but I am also a lover of science—because science is also a search for truth. It might not seem so by today's paradigm, seeing how science sometimes "thinks" in a certain way. . . .

True, but I do honor the free scientists who are breaking through the constraints imposed on them by conventional thinking....

That's true ... and they **are** out there. The first one who pops into my head is Bruce Lipton, who is doing some great work, and there are a few others. But what is really interesting, from my point of view, is that some of the first highly spiritualistic scientists were the creators of quantum physics: Erwin Schrödinger, Werner Heisenberg, for example. These men were very instrumental in creating quantum physics back in the '20s and '30s. Each one of them went on to write highly philosophical books, stating that there is only one mind—only one entity—and we are all part of it. So it is an interesting mix.

You certainly have a profound knowledge of quantum physics—to hear you say you are not a scientist is a little hard to embrace.

I do think of myself as a scientist, but I am sure I am not a scientist in the classic sense. As I said, science is a search for truth and trying to do that objectively, but there really is no objectivity. How traditional science creates objectivity is that they limit the scope of their study, which creates what I call "limited" objectivity.

It also creates "funded" objectivity....

That's correct. I have two sons in college. One is in the PhD program in neuroscience; the other is a theoretical physicist as an undergrad. And well—funding and politics ... that is the name of the game right there.

And it can all get "shmooshed" into a big ball of questionable pursuits! But, again, I do constantly refer to the free scientists, breaking through, discovering new paradigms and finding all kinds of new ways for us to embrace and consider some of the wonderful breakthroughs we are having, and others we are heading toward.

Absolutely.

■ *Let's try to find a central theme to your work—let me ask you to do that. Can you give us a basic idea of what inspires you in your work and what is the thread that binds?*

Two men inspired me to write the book: John Anthony West and Schwaller de Lubicz, a philosopher of the early and middle part of the twentieth century.

■ *... He gave us the masterwork entitled* The Temple in Man—*sacred geometry of the Temple of Luxor.*

Right, and he wrote many other books about Egypt. What he found in Egypt made everything really come together for me—with my own religious upbringing, as well as science, as well as a passion for history and where civilization came from. How he did that was to show that the ancient Egyptians understood metaphysical concepts that are akin to today's philosophy, as it is emerging from quantum physics. So that really struck me. How did these people, so long ago, understand the same principles of quantum physics?

Most likely, they had some sort of programs, technology, educational systems, even meditation techniques that they had been practicing a long time.

■ *Or ... this was an evolved civilization that may have also had additional help from an outside source.*

That's possible too.

■ *Obviously, there is such a spike in history at the onset of the high Egyptian civilization that, from what we have put together and from the clues we have, there has to be more to it than the history books' renditions.*

True, it does seem to appear out of nowhere.

■ *Although, of course, we both are aware of the teachings of Hakim, who described the evolution of Egyptian civilization as having progressed from the Khemit: the prehistoric culture that preceded Egypt. He does not like that theory that there was an outside influence and insists that the greatness of Egypt, as we have it recorded, is the evolution of the Khemitian civilization—the forty-two tribes of Africa.*

Yes, I don't believe it came out of nowhere . . . to be sure. It may appear to be so, but that is obviously not correct.

■ *If only we could get our hands on the books that were burned—that would take us back before the record.*

I agree with Hakim that the origins of Egyptian civilization go way back on the timeline—even so far back as forty or fifty thousand years ago. I think the evidence points to the idea that, at the end of the Ice Age, around 10–12,000 BC, there was obviously a global calamity. We know this for a fact, because we know that a vast majority of species died off and that the species that survived were actually a small part of that.

Particularly from the La Brea Tar Pits in California and another tar pit in South Dakota, we know there was a terrible catastrophe and if man was alive, which we know he was, he also would have succumbed to this calamity.

■ *This coincides of course with the Atlantis theory, which I do subscribe to.*

Exactly. Yes, people keep looking for Atlantis as a city, or an island—but I think it was a vast civilization.

■ *I think it was an entire continent! Having just journeyed from the May-alands in Mexico and Guatemala, as well as Peru—I find so much recurring history and similarities. If you had a continent right in the middle of the Atlantic, you would have access—from as high as Greenland, to the lower part of the South American continent—which would explain so much.*

One guy was putting the center in the middle of the Atlantic, not far from the Caribbean Islands.

■ *I do not subscribe to the idea that Atlantis was an island. I think that the impact of this civilization was so vast that it most likely provided incredible access to explorers and traders. If you imagine a not too distant expanse of ocean between this gigantic continent (which corresponds to the Mid-Atlantic Ridge) and South America, the Caribbean, Egypt, and other places in Africa, it becomes much easier to explain why the same art, myths, belief systems, and architecture can be found on both sides of the Atlantic.*

Right, and it goes as far east as Iran: ancient Persia. There's a river there that flooded and a bunch of graves and stone carvings—beautiful vases and pitchers—were brought up to the surface. People were stealing these sculptures and selling them; the police were having a hard time keeping things together.

Yes, I do believe that there was this huge civilization area from South America to the Nile Valley and all the way into Iran, and that it was decimated by the catastrophe, at around 10,000 BC.

From my perspective, based on the evidence of what happened from between nine thousand years ago and today, I think a major percentage of the population was killed off. Maybe five thousand to twenty thousand people were left: extremely low by our standards today. What's happening between 9,000 BC and 3,000 BC is that civilization is slowly building itself back up all over the world.

Some of the structures that we look at, like the pyramids and some of the great statues in Egypt and in South America, clearly weren't built two thousand or three thousand years ago. They were built more like ten or twenty or thirty thousand years ago.

■ *Hakim puts the date of the Osireion and the Sphinx at 50,000 BC.*

That would probably fit well with the erosion marks on the Sphinx.

In my second book, *Before the Pharaohs,* I did a lot of work with Robert Schoch on the dating of the Sphinx. I've been looking at academic papers written about erosion around the world, and I'm finding now that the Sphinx is extremely old and Hakim may very well be right. There may be scientific evidence to prove that.

■ *Have you been wandering around beneath the Giza Plateau, where they tell us there are no tunnels?*

I have never been down there, no. I have been down a few holes, but that's a closely guarded area, and I don't have the political muster to actually get in.

■ *Right. They say there are no tunnels, of course, but Giza is known as "Rastau," which means "city of tunnels"...*

We sure do know there are tunnels there. We all know.

■ *Getting into them! This is the question.*

That's the hard part—it is so interesting about the tunnels. There is a man up in Washington—a marine engineer, he's actually back-engineered the lower part of the Great Pyramid—who has scientifically proven that the subterranean region of the pyramid was a water pump—a ramp style water pump, that we stopped using in our civilization a couple hundred years ago.

■ *That's right! Christopher Dunn discussed this in the conversation I had with him earlier.*

The pyramid is so mysterious—probably the most bizarre ancient building ever built, with its lower, mid, and upper chambers. It's all granite. And you're right—Chris made a huge breakthrough about ten years ago and he really looked at it hard from a manufacturing point of view. It is extremely plausible, based on how it is put together. But the subterranean area is kind of a mystery. The room down there seems rough and unfinished.

Actually, at about the same time Chris was publishing *The Giza Power Plant,* John Cadman was making models of the Subterranean Chamber, trying to see if it was, in fact, a water pump. He got his ideas from a man named Kunkel, who thought the entire pyramid was a water pump—well, that's not the case. A water pump beneath the pyramid was designed, Cadman found out, for a very specific reason. That reason was to create a shockwave or a pumping action back and forth called a "compression wave" that actually moves up the center of the pyramid. Well, why that?

The top chamber is made of granite, which is the best material you can use for sound. Today, a lot of high-fidelity stereo companies use granite to base their equipment with, because granite has a certain resonance to it that no other material in the world has. For me, what appears to be going on is that this water pump creates a compression wave moving through the pyramid up to the top chamber, and as it moves through the Grand Gallery it is turned into sound, through Helmholtz resonators. It creates two notes that are going out the north and south shafts at the King's Chamber. Why? Recent studies show that a couple of physicists in Russia found out that a pressure wave, which is the same thing as a musical note (as sound is just vibration), would create an electrical field up above the sky. I realize this is really wild....

■ *Not really! Considering HAARP and other secret technologies being utilized today in our atmosphere!*

Right! So we have this electrical field—a bubble of sorts—around the top of the pyramid. Why would the Egyptians do that? One very interesting thing is that this bubble of energy would do one thing that we know of: it would deflect ELF (extremely low frequency) waves that are everywhere around our atmosphere and focus them down to the ground, in various areas.

So, why would they want to do that? Plants are very sensitive to the extremely low and very low frequencies. Studies have shown that, if you subject plants to these frequencies, they actually like it! They actually grow faster and stronger, and have a higher yield.

■ *And I'm wondering if you're going to say something about Agharta— the inner world—and whether these waves are reaching the inner Earth to serve possible life forms there. I mean we're going far out—so why not go a little further? We know that there are complete cities for government and military types, just a few hundred meters down below the surface in all the major urban areas, and that these are partially inhabited year- round—so it's not that "far out" to consider that there may very well be civ- ilizations even deeper: living within the inner Earth.*

Yes, anything's possible ... but that is not where I was headed.

I think that the alignments of the pyramids are north and south, west of the Nile River. You have this engine up at the top, which is the

Great Pyramid, and it is creating a resonance field of electricity. The sound coming out of it is moving in all directions, so that the large pyramids to the south of Giza are picking up this resonance and then creating their own resonance bubbles of electricity... and so fertilizing all the fields—up and down the Nile. Actually, if you think about it, there needs to be a reason for any civilization to build such huge constructions like the pyramids. They are akin to our modern-day power plants—I guess that is a pretty good analogy. I mean, of course, we don't use power plants to fertilize fields....

■ *No, in fact—in many cases, we use them to destroy them!*

I'm sure you will consider that, very possibly, there were other metaphysical purposes for their construction. We have seen how the pyramid is placed at an epicentral grid line, when observing the Earth as the Platonic solid: the dodecahedron. Would you also consider that the pyramid might very well be some sort of activator for the entire earth power or energy lines?

I don't know about the entire Earth, but it would certainly affect the immediate area. Whatever the purpose was, for these civilizations to spend so much effort and so many resources all the way up and down the Nile Valley... I just believe they had to have had a very, very good reason.

John Cadman, who made the models of the Subterranean Chamber, also proved there had to be tunnels under the Giza Plateau, by the water pump model. The proof of this model, which can be tested and replicated by anyone who wants to do it, can be proven scientifically. They can check John Cadman's Web site for more information: www.gizapyramid.com/johncadman1_copy(1).htm.

■ *Having spent some time down there in the Subterranean Chamber, I have to say that you can feel these waves of energy moving through your body . . . you really can. Have you been down there, Ed?*

No, I have never gotten down there. Last time I was there, it was all closed off and you have to know somebody to get in.

■ *I take people down there on my sacred journeys. Next time I go to Egypt, maybe you can be with us. If you want to get into the Subterranean Chamber . . . this I can do for you.*

How? How do you get in there?

■ *Let's say I have friends in high places! It's either that or maybe I haven't been found out yet. . . .*

Hmm. Speaking of that: the second time I was on the plateau, I was out at the Sphinx, taking pictures of the erosion there. A man walked up to me and he started speaking to me nicely, you know, "How are you doing? . . ." He wasn't hassling me, so I started talking to him. Next thing you know he said, "Would you like a tour?" And I said, "Uh, I don't know. . . ."

The next thing you know he pulled out a card and said, "I am with the Department of Antiquities."

He took me places, behind barbed wire, past the guards. It was amazing. He showed me places I have never read or heard about. He took me to one place about one or two hundred yards west of the Sphinx, toward the south. There is the remains of an unknown temple—kind of like a rock shelter—about thirty to forty feet deep by sixty to eighty feet wide.

There are three remnants of granite pillars, fluted on the sides, and carved with hieroglyphics. If you look back into the rock shelter, it goes into a cave area with a huge iron gate that's locked, so no one can get in. For me, this was awesome.

■ *Aha! Something similar happened to me; in fact, I think it might be the same friendly man we're talking about. He asked me, "What is it you're really looking for?" with this probing, deep look into my eyes, and I said: "Everything that I am not supposed to see." And then I added, "but one thing specifically I have come to see is the Osiris Shaft."*

Have you been in there?

No, I have not! I know about it—I've seen pictures of it—it was created a very, very ... very long time ago.

■ *Well, I got the permission to go in and I did. They call it the "Osiris Shaft" because at the bottom of this thirty-meter-deep, straight-down, very scary shaft they've discovered a so-called "tomb," which houses another granite "sarcophagus." It is surrounded by the remnants of four pillars, which is strange and exciting, especially so deep below the plateau.*

Zahi Hawass declared it the tomb of Osiris, stating that they may have found the resting place of the god Osiris—which simply doesn't make sense on any level.

To me, it seems to be part of some sort of electromagnetic device—part of a larger electrical power system—some kind of giant battery perhaps— as I discussed with Christopher Dunn.

Some time ago, I read an account of someone who had gone down into the shaft and said that the tomb is "covered in water—which is more like acid." He claimed that when some pebbles fell into the water beneath him, he could hear them sizzling in the water—which led him to believe that putting his feet into the water would be not too smart. I thought: "There you go....You've got some sort of huge battery network around the pyramid!"

... and there were three granite sarcophagi in there too, right?

■ *Yes, there are two or maybe more at the second level and the one at the bottom—and there appear to be tunnel openings, which are blocked off by debris from the digs. Apparently, according to this writer, it looks like the debris has been placed there deliberately, to camouflage the tunnels.*

I was determined to go all the way down. I got as far as the first tier and thought about how crazy it was—one slip of the foot and I would be history. So I didn't make it that time, but I am determined to go back and go all the way down. I know intuitively that there are tunnels that go under the Sphinx, passing through this area, and reaching the pyramid of Khafre.

When I met again with the man from the Department of Antiquities, who was waiting for me to emerge from the shaft, I told him I couldn't make it all the way down.

I asked him pointedly: "Isn't there an easier way for me to get there?" sure that I could get there through one of the tunnels below.

"We have never acknowledged that there are tunnels under the plateau," he replied, evasively. And I was left to read between the lines.

That is a very interesting concept. There's a certain disconnect between our modern history and the history of ancient times. In his trilogy of books, entitled *Black Athena,* Martin Bernal did a great job of recounting the history of the ancients, versus history as we know it. The history coming from the Arabs, back in the ninth to twelfth centuries, paints a very different picture of Giza. One Arab historian wrote that the remains and the granite debris of the city around Giza extended a half day's walk in each direction. That's a pretty big ancient city. So I wonder what lies below the sands there.

Also, there are writings about a fourth pyramid at Giza, referred to as the "Black Pyramid," because it was either built out of black marble or black granite. It was a smaller pyramid than any of the other three, but supposedly the ancient Egyptians stored their machines and instruments of their civilization in it, almost as if they knew the catastrophe was coming and they started putting things underground, to hide them and protect them ... until the catastrophe passed.

Reisner found the tomb of Queen Nefertari back in the late '20s, early '30s. It was untouched. He held a big celebration of the opening of this tomb and all the dignitaries and VIPs were invited. Ten of these people got down to the tomb and when they opened the sarcophagus, which was still sealed, they found it was empty—very strange. What was important was the other things—furnishing and implements placed in the room. They were so wonderful and so beautiful that today we would have a hard time duplicating the quality.

▧ *Are you talking about scientific tools?*

No, household things—beds, chairs, personal effects ...

▧ *We tend to think of ourselves as the ultimate technologically advanced civilization. But we are often surprised to find proof that the ancients may have had tools and methods superior to ours, as if only contemporary society could have reached some level of high-tech wizardry, which in a certain sense, we have. That is not to say that there have not been thousands of civilizations before us that have reached an apex, and declined.*

That's true. There is a book, *The Mystery of Cathedrals,* written back in the '20s, which says that the secret of the alchemists reveals that humanity has flourished in forty different civilizations that have come and gone over time.

▧ *Back to the plateau. I'm still surprised that you haven't gotten into the Subterranean Chamber. As I said, I do get my groups in there even though it is almost always locked to visitors. They still do unlock it for me. This means that, when I conduct my programs in Egypt, I get private access for them for two full hours inside the Great Pyramid. Some choose to go down to the Subterranean Chamber; others join me in the King's Chamber, where we do some incredible work with sound toning and chanting. Incredible things happen, including Tibetan masters coming through, riding those sound waves. Everyone can hear them.*

The last time we were there, without me ever mentioning the phenomenon, people started telling me they heard Tibetan chants. I have my own theories of how the Tibetan masters travel to the sacred sites of the planet, by riding the sound waves. And of course I'm sure we agree that the Great Pyramid is the quintessential sound chamber of our planet!

Yes, it is!

■ *In the Subterranean Chamber . . . it is just indescribable. I have had a life-altering experience down there. On one occasion with one of my groups, people were toning down in the Subterranean Chamber, others in the Queen's Chamber, and the rest of us were chanting up in the King's Chamber. We began to feel as if the walls were starting to breathe. Many people felt the walls undulating and that they were like multidimensional doorways to another dimension.*

Wow. That is awesome; I would love to experience it.

■ *We talked about the Ice Age and the calamity that undoubtedly occurred. This does lead me to ask you how you perceive what is occurring now—the Earth Changes of our time. I am interested to hear your perceptions about the current global shift and how you feel things are evolving.*

For starters, I have no fear about what is going on now.

There are two different things going on: there's a spiritual experience and a physical one as well.

Physically, what is going on is that the planet is warming up. There really is no doubt about that. Is man the culprit? I don't think so. There's a book out called *Chilling Stars,* in which the author draws a correlation between the temperature of the planet and where we are, in relation to our orbit around the galactic core. If there are a lot of stars in our backyard, as we move through the galaxy, it is going to create excess cosmic rays. As they come through our atmosphere, these excess cosmic rays react with dust particles and create low-level clouds. The more cloud cover we have, at low levels, the cooler the planet will be. The less clouds we have, the warmer the planet; the more clouds we have, the cooler the planet—very simple.

This cycles, goes up and down, over a wide expanse of more than one hundred thousand years. It all depends on where we are in the galaxy. In the last ten thousand years, our planet has moved away from Orion's arm, and it is in an area now, where there are not a whole lot of stars nearby, so that kind of explains the physical part.

The spiritual part has to do with science, on the one hand, and with people seeking truth on the other. We can call it "New Age"—that's a good enough term for it—but people in general are becoming more spiritual. They are awakening to the fact that we are all one abstract entity. We've been fragmented into a physical world, in order to "experience." Some believe there are cycles to civilization, where we enter ages of enlightenment, and then fall back into darkness.

This we were discussing earlier, when we spoke of the rise and fall of ancient civilizations, and so it shouldn't surprise us that the ancients reached great heights in technology as well. . . .

■ *And is that one of the primary themes of your book?*

Yes, the basic concept of the book is that of quantum physics: that we are all One. We are living in an illusion—not a new idea. It is a very old idea that goes way, way back. The actual roots of this ideology go back to the very beginning of ancient Egypt. This belief in the Oneness of all beings is really the foundation of our religions today—Judaism and Christianity, for example.

■ *When you speak of this ideology as "emanating from ancient Egypt"— are you talking about prehistory, back to the Khemit, or are you speaking of the time of the emergence of the Ennead?*

I'm talking about both—I believe the actual origins of Egypt actually go back to thirty or forty thousand years ago.

■ *Okay, great, then we are speaking about the approach of my dear brother, Hakim, the wisdom keeper of the Khemit tradition, who held that it was the ancient Khemitians of Africa who preceded the Egyptians—not the Atlanteans.*

You can actually see this depicted in the gods and goddesses of Egypt—there are several hundred of them. Early on, historians confused their ideas with an animalistic religion, as if they worshipped animals—and that really was not the case. What they were doing was explaining, through symbol, the principles of nature. Actually, it's a science. Take, for example, Anubis: the jackal. Wherever you would see a jackal head on a human form, it would signify the spirit of the jackal or the principle of the jackal, as it applies to human beings.

The principle of Anubis, the jackal, is digestion. The jackal lets his meat rot before he eats it, breaking it down before it enters the stomach. It is a principle of digestion par excellence; he could eat anything he wanted.

They had this system of explaining nature through symbol and the ultimate explanation is found in Luxor in the Temple of Amun.

There we have Amun, Mut, and Khonsu—god, mother, child king. It explains that the nature of man is abstract and it comes from a duality: that man is the Cosmos and that we are all One.

■ *How do you get to that from the Temple of Amun?*

As you walk into the Temple of Amun, it actually tells a story—inscribed in the structure. When you walk in, you are walking into the leg area. And as you walk into the temple, you actually walk through the belly and the chest and into the head.

■ *You're talking about the Temple of Luxor, not Karnak, right?*

Yes!

■ *Okay, so we are talking about Schwaller and the Temple of Man.*

The temple itself describes the physical nature of man, and it also describes the abstract nature of man.

■ *You're suggesting that, as you pass through the various structures of the temple, you are encountering different aspects of the human being?*

Correct. In the chamber of the cosmic marriage, you find the mural, which explains how Amun mates with Mut to create the child king—man. Amun and Mut also substitute Osiris and Isis: they are the cosmic parents of man. What it is saying is that the nature of man is abstract; that he is manifest in the physical body "to experience." This is really the same explanation that we have today, coming out of modern physics.

Getting back to my earlier point, that this idea was, in fact, the inspiration for the Judeo-Christian religions, it got lost. There's also a key point here that I have to discuss: we know that Moses was an

Egyptian—that is very clear. Whether you believe he was real or not, he was said to be Egyptian and he was responsible for writing the first Hebrew Scriptures.

We also know that Christianity was an outgrowth of Judaism, and that Jesus actually spent most of his childhood in Egypt. There's the nativity scene, which is an incredible story in itself, explaining the Hermetic tradition and the Egyptian tradition of Christianity. After that, Jesus goes to Egypt and you don't hear about him again, until he's about thirty years old.

Both the Hebrew Scriptures and the Christian Scriptures have their origins in ancient Egypt. If you're a student of Scripture, and you read both the Old and New Testaments, you can see clearly that the philosophy of ancient Egypt is the same philosophy in the Old and New Testaments and that the Bible, of itself, is not how modern Christianity interprets it to be. It has been turned completely upside down.

The whole idea of the Scriptures is esoteric wisdom—not exoteric fact. Christianity today says that Jesus was an absolute man, he was absolutely God, he absolutely died for your sins, and so on. That is not what it is talking about at all. What the Scriptures talk about—what Jesus is talking about—is that we are born physical, but we don't realize our true nature. There's a search for this "nature," and when you understand the idea that we are all One—that we are the Universe—you begin to identify with that abstract nature of "All There Is."

That is what the Resurrection actually is. It is not meant as the resurrection of the body, but rather the resurrection of your identity of self.

■ *Would you say that the description of the Resurrection in Scripture is an archetypal expression of our multidimensional existence? Jesus was showing us, basically, that we do not die, and that we are souls in eternity.*

Correct.

■ *I think this is the primary lesson in life: that we* don't die. *If you can get a handle on this—your own immortality—it will most likely be one of the greatest spiritual epiphanies of life.*

Yes, that is the core of what he was saying. Specifically, the Crucifixion was symbolic of abstract man becoming flesh—man is crucified as he becomes flesh, because the body dies. However, because man is abstract, the resurrection or realization and acceptance of that abstract and eternal nature is the "coming and return," as Nietzsche put it.

The Crucifixion and Christ's subsequent rise from the dead are not real—they are not literal. They symbolize the search and attainment of truth, and the transcendence of who you are from your own identity—your brain—to the realization that you are the Universe; you are the Cosmos; you are eternal. That's what he was trying to portray.

■ *How does this relate to the spiritual technology of Egypt?*

They got their ideas from ancient Egypt. Ancient Egypt has a lost history. To put Egypt into proper context, that is, the ancient Egypt of four to five thousand years ago, you could compare it to the United States: it was a great country—education, food, culture, growth—people wanted to go there to learn. All the early Greek philosophers wanted to go there to learn.

■ *Indeed. It was truly the epicentral point of culture and spiritual enlightenment on the planet, I would think.*

Yes, so the ideas coming out of Egypt were not only the basis for religions—they were the basis of the Western world. It's been lost, but much has been kept alive as a secret. It was kept alive by Moses. Of course, when he passed away, the ancient Hebrews sort of went astray and then the Christians, Jesus himself, picked it up, and it became the Gnostics who understood that core of truth. Then, after Christianity became an official "political" organization in the third century, it had to go underground again.

Then we get such groups as the Freemasons, keeping this knowledge a secret; this takes us all the way forward to the Renaissance. During the Middle Ages, the Catholic Church had a secret. In that era, they built a lot of cathedrals throughout Europe. There is one near Paris that has all the symbolism built into it, which the Freemasons were not going to talk about: neither were the Gnostics.

This is the heart of the book. A good chunk of it is history, which traces this movement all the way back to the Hermetics of Egypt, through Judaism, then Christianity—up to today.

■ *What I love about your book is that, while you are exploring this evolutionary progression, often very scientifically, you never stray from that beautiful metaphysical understanding of reality, which is why it keeps me glued to it.*

What I am really driving at is this: people want to know who God is. This is the great search. Who am I, who is God, and what is going on? Science mixed with metaphysical wisdom, in the end, we find, are always united; science always points to the big mystery.

There is a number of people trying to define God, others who try to deny God's existence. I know of a man, Richard Dawkins, a scientist, who is coming out with a new book, which states that there is no God. There is more than one book coming out with the concept that there is no God.

Scientists say, "There is no God," because they can't find the evidence, so therefore he doesn't exist. I understand that, but the problem is that they're looking in the wrong place! They're looking for some kind of special sign, which, ironically, Jesus said you cannot find. The punch line is that **we are God.** That may sound blasphemous to some people.

■ *I can't imagine why it should sound blasphemous!*

Jesus said, "We are all children of God."

When people have trouble with that, in a sense, they are not respecting and honoring the Scriptures that they quote, nor the words of Jesus himself.

Agreed. Not that I say, "I am God." That's not correct. A more appropriate way to express that is that we, as a whole, are God. We are all One. For the longest time, I could never really understand why Jesus always said: "Do unto your brother what you would have him do unto you." What was the deeper meaning?

Through my lifetime, I have learned that when you do something to someone else, you are actually doing it to yourself. When you hurt someone, you are hurting yourself.

■ *In this time of extreme polarity on this planet, and the behaviors of separatism that are emerging from the fearmongering that is being choked down people's throats, many are distancing themselves from that sense of Oneness and Unity. But, we are all cells of one body.*

I look at the Universe as a physical body, and we are cells of that body. One gets sick—the entire organism gets sick.

Actually, that metaphor of the Cosmos as a body is excellent. The ancient Egyptians were trying to explain that we are the Cosmos, in that same way.

■ *Right—each cell is a unit of consciousness. In the Temple of Man, we are shown the form of a human being through sacred architecture. Who is to say that the Cosmos does not have that same form?*

Yes, again! We have a habit of looking at things through our physical eyes, and it is engrained in our brains (our sensory organs are physically based) that everything is separate. That's how we understand the world around us. But the truth is the opposite. We are abstractions. This physical thing called the "Cosmos" is actually within us.

There are really two different viewpoints: there's the microcosm and the macrocosm. Man is the Cosmos, and the Cosmos is Man. That is why the ancient Egyptians built all those gargantuan statues, such as those in Luxor, in Memphis, and in Abu Simbel (probably Luxor more than any other place). In Abu Simbel, you find this forty-foot-tall statue of Ramses, for example, stepping forward with one foot. He has a rod in one hand and he has this almost feminine face—round eyes, soft smile. . . . If you didn't see that he had one of those fake beards, you wouldn't really know if he was a man or a woman.

■ *Buddha and Akhenaton are similarly depicted. All of these spiritual idols have the common trait of androgynism: they demonstrate both male and female qualities. To me, as a metaphysician, this demonstrates that they have reached the harmony of male/female balance—the resolution of duality.*

I think it goes even deeper than that. A lot of these statues, like that forty-foot-tall statue of Ramses, have a diminutive two-foot-tall female at his foot. How I interpret that is that the forty-foot statue, itself, is symbolizing Man in the Cosmos (with the capital M), which means humankind evolved, while the diminutive female statue at his foot represents mankind as a whole (with a lowercase m).

■ *Why female?*

Because the female is the mother—the nurturer. She is what really counts. The Gnostic image of Mary shows her holding up an egg. There is a very famous Gnostic portrait of a woman dressed in robes and veils and she holds this egg, which represents the idea that woman is the true source of life—because life is birthed from the woman's body. That is the cradle of life. The diminutive statue at the foot of Ramses is female, because the female is more representative of humankind than the male.

■ *Or at least the birthing of humankind . . .*

Birthing and nurturing. Nurturing is extremely important. That is what propels civilization. That, at least, is my interpretation.

■ *Yes, I wish more people could remember that. Do you think, then, that all of these gigantic statues in Egypt, such as those in Abu Simbel, are depicting the "higher" Man?*

Yes, I do.

■ *Do you dispel the idea that the earlier Egyptians, possibly Atlanteans, might actually have been giants? We do have traces of information that speak of "giants" walking the Earth. When you gaze upon these colossal statues in Egypt, one has to ask: are these representative of a race of actual physical beings, at some earlier time—the Atlanteans perhaps?*

I think the statues themselves are of a superior technology—you can call it Atlantis if you want to (there's so much controversy around that idea). Let's say that the Atlanteans created and manufactured these colossal statues. Were they giants? I don't think they were that tall—but I do think they were large.

■ *Hold on! Are you validating the suggestion that the Atlanteans founded Egypt?*

Yes, that is what I believe. To be more specific, Atlantis was very, very large—not to get into a debate about where the island was located. The whole civilization was huge and Egypt was a part of that. Correct.

■ *I think, too, that there were many cycles of Atlantis, one of the longest existing civilizations of the planet.*

Yes, it did last for thousands of years. If the last incarnation of Atlantis existed from 50,000 BC to 10,000 BC, that is forty thousand years right there! Our civilization, as we know it, has only been around for five thousand years.

■ *... And the United States has only been around for over two hundred years! If you think about what we've created in only one hundred years as far as our technological advancement is concerned ... wow!*

As I recall, you were talking about leaps and bounds of societies and how we can imagine the Egyptians could very well have had advanced technology. If you measure what we have created in a hundred years with the potential of what the ancients could have done in thousands upon thousands of years, it boggles the mind!

We've gone from not flying at all in 1903 to the U.S. landing on the moon in 1969! That is a mere sixty-six-year time span. This is how to understand the context of the pyramids. People still contemplate how they were built by hand, how they dragged the massive stones up by sleds. But if you have been there and seen the pyramids, like you and I have, it is actually ridiculous to think that men wearing diapers put these things together! That's what they show us on these documentaries: all these guys wearing loincloths, towing twenty-ton stones in the broiling heat.

■ *Ha! That is a classic fallacy, to be sure. But speaking of the pyramids . . . you talk about Saqqara in the book. Saqqara is supposedly the first pyramid ever built. But I have a problem with that—how about you?*

Yes, I have a problem with that too. I don't think the step pyramid was the first pyramid ever built. My hunch is that it was a re-creation. What is underneath it and under the surrounding area at Saqqara is something completely different. The step pyramid could easily have been built by men of five thousand years ago.

■ *The interesting thing about Zahi Hawass's take on the development of the pyramids and the Plateau is that the Egyptians kept trying and trying, until they finally mastered the technology, to create the Great Pyramid. I've always thought: "What if those giants of the ancients created the Great Pyramid first, much earlier, and a lot of the others were mere attempts to re-create that mastery of technology at a much later time?"*

I believe there are ten actual pyramids that are true pyramids and all the others are re-creations. The very first pyramid built was at Abu Rawash. That is five miles north from Giza. Have you been? It's a forbidden site.

■ *No, but I will make sure to go there and explore; "forbidden" doesn't stop me! It's such an unpleasant word. Recently, in Abydos, I wanted to see a number of places and the guards kept telling me everything was forbidden. So I went to the Chief of Police and asked him, "What is the point of visiting Egypt if everything is forbidden?"*

We broke bread and I got him to open up some of the sites for me. He planted the seed thought that I could get bitten by a snake or a scorpion, climbing around in some of the more dangerous locations. I promised him I would sign an affidavit that, if I died from a snakebite, I would release the police from all responsibility!

... but I'm digressing. Please tell us how you know that the first pyramid ever built was at Abu Rawash.

I believe the ten large pyramids up and down the Nile Valley are a system of vibrations that creates an electrical field to direct ELF and VLF waves to livestock and crop. The Great Pyramid was the first pyramid built, but it wasn't supposed to be. It was the engine. It had the water pump that created the compression waves, which actually started the process of creating the electrical fields.

■ *You just said it was Abu Rawash that was the first pyramid ever built.*

... but it was never completed. The Great Pyramid was the first pyramid that was actually **completed.** The first pyramid had to be very specifically sited. They built the Great Pyramid on the Giza Plateau because there are rock formations there that create an effect that focuses the electricity underground into the area.

■ *So, the others were built in relationship to the function of the Great Pyramid. I imagine that would explain why the Bent Pyramid is shaped as it is— no doubt that it was created that way to resonate at a certain frequency.*

Right! It also explains why, the farther south you go, the shorter the pyramids become, because you want a diminishing negative electron flow coming off the pyramids, to help pull down electrons from the north, creating this field. There's a whole science involved in this that will be in my next book.

Back to Abu Rawash. They started building at Abu Rawash, cutting the pit at the bottom, and they got about fifteen layers of stone going before they stopped and left everything. Why? Because they found a better site, five miles south at Giza, which was more suitable for the engine. That is how I explain it.

Abu Rawash is fantastic—you have to go there. No one is there, so no one will bother you while you explore the site. It's just you and your guide.

■ *What am I going to experience there?*

You're going to experience an unfinished pyramid, but I tell you—when I was there, the spirit of the ancient builders was there . . . more than anywhere else.

There's another surprise there. Chris Dunn actually found it. On the south side, there is a stone about four feet long by three feet wide, about eighteen inches thick—a slab of red granite. It was cut with a forty-foot circular saw. You can actually see the feed lines of the blade, and you can see where the blade stops—it makes a primitive arch. It is the most beautiful thing I have ever seen. And it is absolute proof that whoever built the pyramid did so with power tools. There's no question about it.

We live in a paradigm that says what history has to be, but we are learning it doesn't have to fit it.

And then there's Dendera—have you been into the crypts?

■ *Yes.*

Did you see anything unusual there?

■ *Yes, there are found the exciting images of what appears to be two artificially powered light devices—what look like large glass tubes and a filament running through the center, like our modern-day light bulbs.*

Yes, and the images of the people carved—their headdresses seem very unusual, don't they? These crypts are corridors. Actually there are three of them—but two are closed to the public. I think they are a Hall of Records. I would like to know what is in those other two crypts, and why they are closed to the public.

I know you have your way of getting into these places, so, if you do, take pictures of every single mural and image. Some of these are very, very strange.

■ *Well, most of the glyphs of Egypt are strange, according to our modern understanding, don't you think?*

Yes, but those in Dendera are **extremely** strange. I only know of one man who has tried to tackle the project of deciphering them. Doesn't that tell you something? Why isn't there more information available on Dendera? Why aren't these murals made public? Why aren't they described, or explained?

It's the same as the "unknown temple" on the Giza Plateau that I have been to. I had no idea it was there—never a picture . . . nothing.

■ *I'm still not sure what temple you are talking about—that's how unknown it is.*

It's about one hundred yards due west of the Sphinx, at the southwestern edge of the Plateau. It's built in a rock shelter. There are four of what is left of columns, of red granite—huge—they're monstrous. I couldn't believe it when I saw it.

■ *What about the Temple of Isis at Giza—did you see it?*

Temple of Isis??

▤ *Yes, the person showing me around told me that most people were unaware of it. We're just not allowed to go to these little corners of the Plateau.*

Now they're putting surveillance cameras and fences all around the Plateau, and I believe soon we won't be able to go there at all. This is why I encourage people to get to Egypt as soon as they can; it is why I continue to bring spiritual groups there.

It is such a shame. I mean, yes, the Egyptian government is in control of their antiquities, but I think that the whole of Egypt is, that it is the cradle of civilization for the entire world. They have a world treasure there.

▤ *I just can't help thinking that, on some level, they are afraid that we're going to activate the energies of the sacred sites. It is happening everywhere. In Peru they're shutting down the sites. Pretty soon people won't be able to enter Machu Picchu. In Mexico, they're fencing in all the sacred sites, blocking people from free access, even the natives. Why is this happening? When I was in Mexico at Chichén Itzá, there had been satellite trucks and NASA equipment there shortly before me, and the story is that George Bush went into the pyramid with Vicente Fox. Apparently they went up to the top of the pyramid and, immediately after that, it was forbidden for people to go up.*

Yes, something is going on—I do not like it.

▤ *I don't like it either! Let us just remind people that the sacred sites, with all their mystery and magic, are our heritage and that, as spiritual beings, we have a calling to bring the love and energy there, working with those vibrations that the ancients knew how to encode there. They built them to last—to make it into this moment, perhaps, when time as we know it ends,*

helping us shift into that higher awareness that we are beginning to real-
ize is the legacy of humankind.

■ ■ ■ ■ ■

Edward Malkowski is the author of:
  The Spiritual Technology of Ancient Egypt
  Before the Pharaohs
  Sons of God, Daughters of Men

For more investigations into his work and discoveries,
visit his Web site at:
  www.sonsofgod-daughtersofmen.com

José Argüelles
and the Galactic Maya

■ ■ ■ ■ ■

Stephanie South
with José Argüelles

Stephanie South is a former journalist and co-author of the seven-volume *Cosmic History Chronicles,* of which the first four have been published. She is currently working on the fifth volume, *Book of the Time-Space.* She met José Argüelles in 1998 after a series of dreams in which he demonstrated time travel and telepathy. She is currently working with him on the Noosphere II Project, an investigation into the nature of time and cosmic states of consciousness. Her book *2012: Biography of a Time Traveler* is not only an engaging journey through the inner and outer realms of the visionary José Argüelles. It also sheds light on the crisis our planet is undergoing today and offers clues about how we can realign with natural time, to make a peaceful transition into the next cycle of consciousness, which begins December 21, 2012.

■ ■ ■ ■ ■

▧ *The book is such a temple of love and respect for José Argüelles and his contribution to our evolutionary experience. How did you meet and become involved with him?*

I first met José at the 28th Annual Whole Earth Festival in Davis, California, which he had started in 1970, when he was a college professor. Before that, I hadn't heard much about him—I had heard a little bit about Harmonic Convergence, but I really didn't know exactly who José was. Before that, I'd had several dreams in which a man was demonstrating telepathy and time travel to me, and when I first heard José speak and I met him, I recognized him as the man in the dreams. The first time I heard him speak at the festival, he was talking about how the world is living in artificial time.

Something about those words really woke me up!

▧ *The concept of time and timelessness is one of the preliminary themes of my work as well, so I am particularly interested in that aspect of your book—but, of course, it is only one of so many threads in the book.*

Indeed. José's works are so vast that I felt that it somehow had to be compiled into some sort of a template, in order that other people could understand the different experiences and the teachings that he has brought through. What most people remember above all, of course, is Harmonic Convergence in 1987.

▧ *Yes, that is the first I heard of José Argüelles myself. Can you talk about that experience? He actually organized the global event by awakening people to this important time on the planet.*

Yes, that's right. Harmonic Convergence occurred August 16–17, 1987. He had been given those dates in the early 1970s: those were the prophetic dates of the closing of a certain cycle of Quetzalcoatl, but he didn't realize then that it would lead to the Harmonic Convergence, which would mark the first time that human beings around the world would come together, in synchronicity, whether through prayer or whether actually uniting at sacred sites. Many didn't even know what was pulling them there. They coordinated their prayers and meditations and ceremonies all over the planet.

There was some kind of "magnetic pull" that many people described feeling, which resulted in millions of people simultaneously coordinating all their prayers on behalf of the planet. José had been given this date, as I said, in the early 1970s and this was basically the fulfillment of a prophecy that opened the portals to 2012.

■ *When you say he was "given" this information, are you referring to his spiritual guides? Was he connecting with the Galactic Maya?*

Yes, he was connecting with the Galactic Maya and, in the early 1970s, he met a man named Tony Shearer, who was a Native American poet and storyteller. Tony had been given those dates by a female shaman some years earlier. He passed those dates on to José—it was the closing of the prophecy of Quetzalcoatl, the Mexican mythic culture-bringer, a huge figure in ancient Mexican legend. He received these dates from Tony Shearer and then, in 1983, José had a really powerful visionary experience when he "saw" the Harmonic Convergence unfold. He saw people gathered in circles all around the planet, lying down, with their heads pointed toward a fire in the center of the circle, and he kept hearing the words "Earth Surrender Rites." He immediately knew that this was related to the Harmonic Convergence and he realized that he had to do something to pull people together at that time.

■ *Insofar as his visions of these synchronistic moments on the planet are concerned, how does he interpret the importance of Harmonic Convergence in relation to 2012?*

Harmonic Convergence was one of the prophecies—the prophecy opens from August 16, 1987, to 2012 and basically José feels that 2012 is the **big** Harmonic Convergence.

■ *There's so much expectation about this date, December 21, 2012, as you are well aware. People desire profound information about the prophecies of the Maya with regard to this timeline. I would certainly consider information from José Argüelles "profound."*

That is true. There is so much out there now, and a lot of it is doomsday material.

■ *Right—this is why I do my very best to help people understand that the Mayan vision of this time frame was that it marked the end of a Great Cycle of Time and the beginning of a new one, and not the End Times— as many fear. People seem to have great difficulty grasping even the abstraction of death as the End, which we know it is not, from so many who have contact with the other side, or come back from it.*
Even in the tarot, the Death card indicates transition—the passing from one form to another, although it is often misinterpreted to read the death— the physical death—of a person.
How does José describe this New Cycle that we will be entering at the end of our calendar year 2012?

To begin with, José sees 2012 as the shifting of timing frequencies. Basically, we are leaving one time cycle and entering into a whole other cycle, which he perceives as being one of spiritual consciousness exploration, as opposed to our current materialistic or dense focus.

■ *How do you summarize* The Mayan Factor, *José's "breakthrough book," and the importance of that book?*

All of his works are based on three primary elements: time, synchronicity, and art. *The Mayan Factor* really paints a picture of how the current cycle is closing and the new one is opening, and it shows that the cycle from the beginning of written history in 3113 BC to 2012 was the cycle of "time is money" and male dominance. It was one huge cycle that played out to see what kind of civilization humans could create. Now we're entering another cycle, beginning in 2012/2013.

José has strong connections with the Galactic Maya. He has many encounters with them, many experiences. They are monitoring all the affairs of our three-dimensional reality, and they are watching this whole program play out. The program we are living out now has been foreseen by ancient "seers" long, long ago.

■ *Would you enlighten us to exactly who you intend are the "Galactic Maya"?*

According to José's works, the Galactic Maya are basically star travelers ... time travelers who incarnated here, at a specific time, to seed the planet and to leave clues of our planetary being for this stage of our transformation, when, basically, we would reconnect with those higher dimensional beings—at 2012/2013. They left glyphs and clues, which José has basically spent his entire life studying and decoding and which have provided information about the Big Shift ... about the return of the Maya. The Galactic Maya were masters of time, time travel, telepathy, all of those supernatural powers that we were actually supposed to embody. It's just that most of us have been cut off from these paranormal powers, due to living in a world that is limited to artificial time.

■ Would you define "artificial time" as being the essential aspect of three-dimensional reality? I mean, I think where we are headed here is that we are moving into the next dimension—the fourth. In essence, we're moving out of artificial time and its limitations.

Yes, right. We are moving from the third to the fourth dimension. Now, on many levels, people are really space-oriented. For many, what they see is what exists, yet we know that there is a whole other story playing out, beyond the matrix of matter. So, basically, what's coming into play more and more in human consciousness is the quest to know what lies beyond this three-dimensional world of illusion and appearance. We are becoming more conscious of the fact that we are co-creators of this particular reality that we are all involved in.

■ Yes, and when you speak of the world that we "see"—we need to remember that it is the world that we "think" we see. We do actually see a lot more than appearances presented to us as "physical reality." If twenty people observe something, there will be twenty different experiences of it, depending upon the evolutionary and spiritual development of each individual. I do like to remind people that even the stars that we see in the sky are quite possibly no longer there. That is a perfect demonstration of illusion.

That's really true. It's almost like we're all living some strange memory!

■ Right. And let's think about that for a minute: a star that is two hundred thousand light-years away from the Earth could very well have burned out years ago, but the light still appears, due to the time it takes light to travel the distance. It really helps people grasp the illusory nature of reality when they are presented with the idea that the time element of our current perception can cause things to appear which are actually no longer present ... if that isn't illusion, then I don't know what is!

Exactly. And here we are, almost seven billion people with seven billion different perceptions of reality....

■ *Well, like José, most of my life I have known my connection to the stars; it has been the most significant force in my life, to be sure.*

Stephanie, in the book you speak of the "AA Midway Station" as being part of the galactic federation. I had never heard this discussed, prior to reading it in the book, nor have I encountered it in my own work with the Sirian guides with whom I am connected. Can you elaborate upon this for us?

The AA Midway Station is a type of monitoring base between the fourth and fifth dimensions, according to José's experience. What he saw there was a system of monitoring screens with simultaneous views of what was happening on this planet and others ... and other star systems. It's kind of like the central control system that is overseeing the timing of the entire universal program.

He understood that there is a complete record of all activities—maybe we call it the Akashic Record—of all events ... of everything that occurs. This AA Midway Station is the place where all this information is stored and all the higher intelligence is surveying all of this higher, cosmic unfolding of what we're going through. He witnessed different aspects of the unfolding, and saw how important it was that the Earth was undergoing its revolutionary process.

■ *Hmm. As much as I love the idea that there are higher beings overseeing the developments of all planetary bodies, it bothers me to think that there are even more giant surveillance screens out there. I have my own ideas about the real reason behind this extreme upsurge in surveillance and monitoring of all beings on this planet, just now—at the time when we are about to go through this shift.*

Well, what he experienced at the AA Midway Station was not like we perceive monitoring being done on Earth. It has a more invasive element here on Earth; at least that appears to be how we think about it. The way that he experienced it was magical. I suppose the expression of it as "surveillance" is not the right wording. What he observed was that we are headed toward major synchronization, and that our current program is like a time-release program. It has everything to do with synchronicity, with our release from artificial time, and turning the Earth back to a work of art.

■ *It is so refreshing to be able to provide people with this perspective, and from someone as grounded and at the same time empowered with such incredible gifts as José embodies. He has the credibility to bring this in with the academic knowledge of his experience, and yet through the eyes of a mystic. Tying these two aspects together is truly an art. He gives us a very detailed perspective on the true depth of the Mayan wisdom and how they were empowered to receive prophecy of what lay ahead for humankind: a positive shift for the Earth.*

One other element of *The Mayan Factor* describes how, in 2012, we are basically passing out of a particular galactic beam. The galactic beam represents a cycle of history. If you could just imagine that there are different galactic beams: for example, at one time there were dinosaurs. They became extinct almost overnight.

We're passing to another galactic beam—another time. This other time, on some level, I think, is our collective dream, which is why it is so important that we're putting out positive visions right now of what we all want to create of the next world.

■ *Indeed. And I think that fact that we are being squeezed to death with every kind of doomsday scenario—the economy, the warrior world—we are, in a sense, being forced to let go of everything we believe provided our security and our safety. We are being compelled to look to the higher life values that are pretty much the polar reflection of these darker aspects that are predominant today. People are also being made to realize that they don't need the huge house, the big salary, the high-end computer toys and distractions to survive.*

Like you said, we're going through a huge planetary and personal initiation. Most people can feel it is a time of enormous transition. How that's perceived, according to the 2012 prophecy, is that when you come to the end of a cycle, all the accumulated karma from all of history, all of the lost world, gets brought up to the present moment . . . to be ultimately resolved and released.

■ *Wouldn't you agree that whatever karmic debt we have as a planetary civilization, we will still take with us (to some degree) into the next dimension? I don't believe it will be automatically released, as some people interpret this shift. I think that, faced with this karmic overload through all time of earth consciousness, we will be given the opportunity to raise the consciousness of the planet.*

In shifting to the next dimension, I don't believe we have instant release from that karma—the "utopia" scenario—but rather, we will most likely face every layer of karma that has been created here and be given a chance to heal it, or raise it, as part of our process. In essence then, we bring into the next galactic cycle—or "beam," as you describe it—what we have resolved or not resolved of our karmic bag. Do you agree?

I understand what you are saying. I think the full palette is available to us, and anything could happen. It could be that; it could be that on some level the galactic "refresh" button gets pushed and everything becomes new . . . or it could be that we do have to carry some of this. I think the experience will be determined by our individual consciousness.

■ *Yes, again we are in agreement. All possibilities lie before us. And what about the consciousness of the celestial beings, so significant to the Maya? Let's talk about the sun. You discuss the sun and how it emits patterns of energy, which, again, is very synchronistic with the information I receive.*

My understanding is that the sun, a conscious being, knows exactly what it is doing, working to purify and prepare the Earth and the other planets in our system for this transition. How do these patterns coincide with the Tzolkin calendar?

The Tzolkin is the 260-day count. José saw that, basically, the Tzolkin correlates to the binary sunspot cycles: that, basically, the sun is operating through a galactic program. Since all life in the solar system comes from the sun, all of the solar programs that we're receiving affect all life. The sun is continuously emitting different patterns of energy and information, but the binary sunspot cycles are twenty-three-year cycles. The next one peaks in 2012, which is pretty interesting.

■ *That is remarkable. We have the end of a twenty-three-year sunspot cycle and the eleven-year cycle, calculated in more contemporary calculations of astro events, both coinciding in 2012. 2012/2013 marks the end of a solar cycle, as well as the end of the Great Cycle of Time. It will be very interesting to see what the sun is doing during this very specific time reference point. I think we are in for some real fireworks!*

In fact! We have reports that at the moment there is zero sunspot activity—that it is abnormally quiet at the moment. Some people believe that it is because the sun is getting ready for its peak in three years. There are most definitely a lot of different theories of what is going to happen when that occurs. NASA has issued a lot of reports about the potential of solar peak—for example, that a solar flare could hit and blow up the electromagnetic grid. There are all sorts of theories about how this is going to evolve and unfold.

■ ... *and, of course, the government tends to present the negative aspects of what can happen, so much doom and gloom: asteroids hitting the Earth; the sun disintegrating the Earth. . . . I hold to the belief that the sun is a conscious, living being—it knows exactly what it is doing. And again, all the planetary bodies are conscious beings.*

If we do believe that we are, in fact, entering a New Cycle of heightened awareness, and if we embrace the knowledge of the Maya, then, by nature of those beliefs, we would have to embrace the idea that the sun has awareness of its role in the Cosmos, and knows what it has to do. It is simply part of the galactic scheme of things.

Exactly. It's an interesting contemplation, for example, to consider how fragile our systems are—like the Internet and our communication networks. An earthquake can even bring the Net down. I don't think that this is necessarily a bad thing. Seeing how, on one hand, the Internet is like a form of virtual telepathy, it may be showing us that now is the time to really cultivate the mind powers of telepathic communication. Ultimately, we should reach a stage whereby we can communicate directly with anyone we need to, utilizing our capacity for telepathic connection.

■ *I agree. That may very well be why the Net may have to eventually come down. We do need to trust that we have these capacities, and start to use them around the globe. I think people are awakening to their own innate abilities, with so much work being done on their own spiritual soulquesting. There really is an amazing increase in the number of people who are awakening now.*

Yes. It's happening rapidly; it feels like whatever anyone thinks about 2012 . . . on some level it seems to serve as a trigger in our DNA. So many people who may not have been interested in so-called spiritual things are being activated, in a sense, just by the approaching of the date. It is definitely creating some kind of awakening for people, whether they choose to go negative or positive with it.

I don't know if you know it, but there is a huge 2012 disaster movie coming out in November: a huge Hollywood, big-budget movie. It's kind of interesting, because people in smaller towns and those out of the loop may not have heard about the 2012 paradigm. It is just unfortunate that their first take on it will very likely be one provided by this big disaster film.

■ *But then, that is the mandate that the media have to follow, don't you think? If you keep the masses in fear and ignorance, you can control them more easily. I think the controllers of this planet just wouldn't know what to do with a civilization of seven billion freethinking people. Better to keep them in poverty, in hunger, in fear, in slavery—and entrain them to think that only the government has the solution. That's control.*

Exactly! And it's definitely all playing out right now.

■ *It's interesting sitting here, watching it, I'll tell you. Still, I'm utterly convinced that we are moving rapidly toward a Great Awakening, and nothing can stop us.*

Absolutely. That is what makes it such an exciting time to be alive now.

■ *Speaking of exciting things unfolding right now, there is a lot of discussion about disclosure of the ET connection and, obviously, José enjoys direct contact. Can you share with us what he feels is the ET connection, in relationship with the unfolding experiences that await us at 2012?*

Well, first of all, José feels that extraterrestrials are misunderstood by the masses, in terms of how everyone perceives them. He has had a lot of what you might call "contact" over the years. He's really felt that, basically, the intelligence that he's been in contact with led to the calling for the Harmonic Convergence and to other things that he's done throughout his life, which has basically been channeled from higher-dimensional beings.

Going back to what we were discussing before (I know we don't like the word "surveillance"), basically about beings watching over or guiding the whole evolutionary unfolding of events on Earth—and the UFOs that are being spotted, José views the UFOs as interdimensional, Earth-generated, galactically programmed "information bearers" for us.

There is a lot of interest in extraterrestrials now, but he is careful not to bring the focus on that aspect, just because people tend to sensationalize it all. But yet, he definitely has been (and still is) in contact with higher-dimensional entities.

■ *I think it is of the essence that people understand the difference between "extraterrestrial" and "extra-dimensional." As you know, I, too, am a channel for these extra-dimensional beings. I think it is realistic to consider that there are three-dimensional, physical beings out there, flying around in ships. Of course, there is a perennial debate about whether or not the ships being sighted are just holograms, or whether they are real. There sure is a lot of buzz around this, right now. I can't help but wonder if we are going to have a breakthrough and discovery right around the time of 2012 . . . indeed, this is what I have been given from the guides I connect with. I do believe what I hear and I personally believe we are so close now.*

This leads me to my next question. What do you think of the crop circles? Have you or José entered any of these formations?

No, I haven't been in, and neither has José. He's seen a number of the images, and he definitely sees them as part of the signs of the next cycle.

■ *I will be glad to send you images and information if you would like. What about the so-called "Face" on Mars? You do mention this in the book, saying that it held a great significance to José. Can you discuss with us what the Face represents to him?*

He was shown the Face on Mars by [Richard] Hoagland, who many people know of; he used to work with NASA. He was shown the Face on Mars in 1983, and when he saw it, it served as a kind of trigger for his memory—of times on other planets, or on other worlds, and it really blasted him open! Four days after he saw the Face, he had the huge vision regarding the Harmonic Convergence. The minute he saw it, it served as one of those "wake-up calls"—you know, those moments that accelerate your consciousness, or your evolution.

■ *You mention that the Arcturians and the Antareans are keeping an eye on Mars. Can we discuss what these beings are watching for? Do you believe that there is a civilization currently on Mars now—or is it earth beings, humans, who have burrowed in, under the surface, and are colonizing there?*

He basically feels that the Arcturians are monitoring this, and says that there is a civilization on Mars. Preceding Mars was the civilization on the planet Maldek, which is now known as the "asteroid belt." The story that he received from the Arcturians was that, first, there was life on Maldek—that was the original Garden of Eden. Something occurred, some kind of dissonance, and the planet actually blew up, resulting in the asteroid belt. All of the karma from that lost planet was transferred to Mars, which created a civilization that became extinct. All of the karma from both of those planets and their civilizations has been transferred to the Earth.

■ *Whoosh! That's a lot of karma! It kind of validates my original suggestion that we may very well be taking our karma along with us to the next dimension.*

Yes, the experiences that we often go through that we can't place, like déjà vu and other perceptions, may very well represent other states of consciousness—where we were in other worlds, perhaps. Maybe we're back to hopefully try to correct some of that karma and raise the vibration of this planet, as best we can, during this incarnation.

■ *Michio Kaku, the master of theoretical physics, speaks of how the many levels of dimensions are so close to us—they could be right under our nose! Fascinating to see our leading physicists and scientists trying to figure out how we shift into these other dimensional realities. No question that déjà vu could very easily be explained as our slipping into that other dimension where the activity or events that you are experiencing have already played out—in a nanosecond from the moment you are experiencing it.*

Indeed. It seems that people are experiencing those moments more and more. A lot of people talk about the 11:11 time prompt, and the flashes of other dimensions that happen so fast you can't fully grasp them or remember them . . . but you wish that you could!

■ *For sure! I can personally attest to that!*
    . . . back on the question of ET: I do like to remind people who say they don't believe in extraterrestrials that we are currently running robots on Martian terrain. These are extraterrestrials of human creation (without getting into the debate of whether or not there are also human beings in the underground, as many believe!). People don't stop to think about it, but

we have had two or three robotic craft on the Martian surface now for years! And that is only what we are allowed to know about. What are they doing there? Are we really to believe that all they are doing is photographing rocks and taking samples of the soil?

So right! I had an experience for the very first time in 2000, at Palenque, in Chiapas, Mexico. The night we arrived there (and I had never experienced a UFO before then), there was a group of about eight of us—the day before we actually had planned to go to the Temple of Inscriptions. The sky was filled with UFOs; there was just no doubt about it. We looked up at the night sky and actually saw a huge mother ship! A couple of people in the group were ufologists and they had never even seen anything like it. The feelings that came with the experience were indescribable, so I'm definitely convinced.

■ *Ah! Palenque. This is such an amazing connection. I'm just back from there, as I work with the shamans there. In fact, during this last visit, I was embraced into the family of the Lacandón tribe and appointed as one of the four spirit guardians of Palenque. The Lacandón are, as you probably know, the only remaining direct descendants of Pacal Votan. The spiritual guardians of Palenque, with whom I have been blessed to share in this great honor, are Kayum (the primary guardian of Palenque), Antonio Viejo (one of the Elders of the tribe), and Hunbatz Men (the well-known Mayan Elder of the Yucatán). It is an incredible honor for me and also for the female energy which has been empowered through this.*

They are amazing people. That is so remarkable, Patricia—it's absolutely fascinating.

■ *Thank you—yes, it is! So, I know where I am going to be in 2012!*

As you know, José's work is based on the prophecy of Pacal Votan. That is the origin point of so much of it.

■ *Did you manage to get into the tomb of Pacal?*

Yes, I did.

■ *So did I! We can't divulge how that was arranged, as it is not "allowed," so I am sure neither one of us wants to put in writing the name of the person who let us enter, out of concern for the individual in question. It was quite a powerful experience . . . a very deep, very emotional experience that defies description. I sat for a long time afterward, at the top of the Temple of Inscriptions, with only the moonlight illuminating the grounds of the Palenque site, way down below me, and the sounds of a jaguar in the jungle, just behind me. It was memorable. But I didn't see a mother ship—that probably would have been the moment when I left Earth altogether. What magic!*

Yes, it was. I've never seen anything like it since. And there was no question about it . . . about what it was. We did have some skeptics in our group—not even they could deny what they saw that night. Like I said, it was a huge mother ship . . . that's the only way to describe it.

■ *I imagine it was indeed breakthrough information for your skeptics, to stand in the grounds of Palenque and see a mother ship overhead! Welcome to a new world of perception!*

Through all your learning experiences with José Argüelles, what has come through on the big screen of this most important shift . . . that is, how do you advise people to help them prepare and understand that it is not necessarily a finite date—and that the process is now, just as it will be when we proceed from it—into the next dimension?

I do believe that the shift has already begun and we're already going through it. José feels that we basically have to change our relationship with time: what time is; what time travel is. We need to explore time just like space—the exploration of the material or horizontal plane. He thinks of the fourth dimension as time—the vertical plane.

After so many years of study and experimenting with different cycles of time, he found that living by the thirteen-moon, twenty-eight-day calendar, which is based on the Mayan Tzolkin (the 260-day count, meshed with the 365-day solar count), that it absolutely increases your levels of consciousness, and basically attunes you into a whole other level of reality ... just by following different cycles of time. At the height of the Mayan civilization, they were working with seventeen calendars at a time. They didn't view time as we do. They mapped the night stars and synchronized different events with different star systems.

I've been using the thirteen-moon, twenty-eight-day calendar since 1998, and I can definitely attest to how it shoots your frequency out of the twelve-month Gregorian calendar, and the constraints of the sixty-minute clock! You really have to try it to feel the effects that it has on your consciousness. It is profound.

■ *In fact, I don't wear a watch. It's about fifteen years or so—they don't work on me, anyway. People are usually blown away that I can usually tell the time to within minutes. When they ask how it is possible, I tell them that it is a matter of tuning in, and simply becoming more aware of the phases of the sun and the moon, and just developing a sense of time telepathically. You become so much more in tune with the natural progression of things ... of nature.*

And if you combine that with living in the different cycles ...

You know, being involved with Sirius through the Sirian revelations, that July 26th is the heliacal rising of Sirius in the night sky. The thirteen-moon calendar begins then—the Egyptians used a similar calendar and their New Year's Day was July 26th, based on fifty-two-year cycles. Living in this different cycle, your mind gets pulled from the Gregorian grid, programmed with different holidays and weekends, etc. You can still plug into the thirteen-moon calendar, if you wish, but it is like using a new lens to view reality.

■ *Does your book talk in more detail about this calendar?*

It does, yes. It talks about how José discovered that and then began living in different cycles of time, like the thirteen-moon cycle—that's a big one. Actually the calendar is going to be available soon: I am currently creating the *Thirteen Moon Star Traveler's Almanac of Synchronicity*, based on José's work, which will hopefully help people understand about living in different time cycles.

■ *That is going to be a great gift to people, and so important for those who are experiencing time shifting now. It seems to be one commonality amongst all people, even those who are dedicatedly not in pursuit of spiritual things. Even the die-hard pragmatists talk about how time is not what it used to be, and how there are not enough hours in the day ... all experiencing it on some level or another. People do feel they are losing time.*

So, as we are running out of time now, is there a last message you would like to share?

I think that now, more than ever before, is the time to discover your inner calling: what your inner gifts are and how to create your art—what you have to bring to the planet. We need to do everything that we can to lift the vibration here. The planet needs all of our love and all of our light and all of our creation. I think also that it is important that we begin to explore the nature of time, and so I recommend that people find out about the Gregorian calendar that they use—what it actually is—and to begin experimenting with other time cycles. I guarantee you, it will change your consciousness and raise your frequency.

I encourage everyone to read the book, not just because I did it, but because José has so many different messages, so relevant to the time we are living in. These hopefully come through in this book. I try to tell the story of his life, but also try to give the clues to the times we are living in and hopefully what we can do to make it through this transition.

■ *What better way to approach a dimension, where time does not exist, than to start practicing how it feels to let go of it? Remember, we are already in the shift, exploring a whole new way to be residents of this great planet, on our way into a new world.*

Stephanie, my love to you and to José. Please tell him what a great joy it was if just to brush up close to him, through you, in this daring conversation, from somewhere just beyond the matrix of time.

■ ■ ■ ■ ■

Find more about Stephanie's works, and her organization, the Galactic Research Institute, by going to her Web site: www.lawoftime.org.

The End of Time

Drunvalo Melchizedek

As always, I am so honored to be able to bring you some of the great minds that are helping people break out of the box of convention and soar with the eagles in the free-thought zone, where science and spirit embrace. Helping us do that is Drunvalo Melchizedek, renowned the world over for his spiritual teachings of the Flower of Life and the Merkaba, the self-declared "voice" of the Mayan leaders, who return to a place of leadership now in this transition that they call the "end of time."

■ *Dear Drunvalo, thank you for this very special time together. We're going to be talking about the information in your latest best-selling book,* Serpent of Light, *which describes the Great Shift in the kundalini energy of the Earth. It is so important to contemplate how Earth is a living, conscious being and how, like us, she experiences energy activations and shifts. In previous conversations with other avant-garde thinkers, we have considered man's attempts to harness that energy!*

This coincides with your teachings and with mine, as well as the beliefs of many other master teachers, as we make our entrance into the 2012 paradigm. It is important to remember that no date is fixed in stone and that we don't have to limit ourselves to the date itself. What matters is the transitional process we are passing through, and we can feel that that has already begun.

Gaia has her energy vortices, many of which are home to sacred sites and ancient temples, and so many of us are taking ourselves to these locations now. We go there to receive information and energy, but more—we go there to bring the power of love and heightened energies and healing into her being. You are certainly doing that through your journeys throughout this great planet and, together with my groups of seekers, I too am serving in this capacity.

With all that is in this book, I wonder: can you provide us with an overriding theme of this material? How did it come about?

For a long time, I was not allowed to talk about it in any way. After many years, at about 2006, my spiritual teachers, who are indigenous people, told me I should write about the concepts that I have presented in the book. Actually, there are thousands of people on the Earth who are aware of this information, but they are mostly indigenous people. They told me to bring it forward, and I did.

■ *I know you are very connected to the Mayans, as am I, so when you speak of the "indigenous people," are you talking about tribes all over the planet?*

Yes, there are others that are at least as important as the Mayans. Some people are not even aware of it, but they are vital to everything that is going on around the planet, which is bringing this new consciousness to the Earth. Without the indigenous people, we simply would not be able to do this. We do not have the consciousness, at this time, to do it. We need them to do this and we also need them in ways that hardly anyone understands.

Without them, the whole Earth would die—we wouldn't make it. Our consciousness is not sustainable.

■ *Yes, we are so detached from the Earth as a species. The indigenous people have been able to hold the Earth, hold the vibration, and hold the memory through prayer . . . sound—things that we are relearning now.*

They have memories that go back a very long ways. The Mayan councils say that the Mayans can remember everything going back a little more than twenty-six thousand years, so they've been here at least that long. The Waitaha can remember back sixty thousand years. Some of the Elders can actually repeat the names of their entire lineage, from the great ancestors to present day, all the way from sixty thousand years ago.

■ *I remember when we spoke earlier and you told me that. It was beyond my comprehension how they can possibly retain the memory of sixty thousand years of their ancestors by name!*

And those names are really long. . . .

■ *Is it true they had longer life spans way back then? Perhaps that is how they managed to accomplish such an incredible feat!*

Yes, they did have longer life spans. We keep dropping lower, even though we think we're getting better. Most of my family has enjoyed life spans over one hundred years old, except for one, who made it to ninety-seven. All the others in my lineage have reached one hundred ten, even one hundred twenty, years old.

■ *That's good! We're going to have you around until 2012 and way beyond!*

You said at the beginning of our conversation that they did not allow you to bring this information forward, but that now they have said you can. Why now?

This is interesting. I wrote the book and I turned it over to my publisher in June 2007. In July 2007 the head of the Mayan Nation, Don Alejandro Cirilo Perez, came to Sedona, where I live, and asked for a meeting. To explain who he is: there are four hundred and forty tribes in the Mayan Nation in Mexico, Belize, and Guatemala, of which each has an Elder that has been elected to lead his tribe. There are always four hundred and forty elected Elders, who are members of the National Mayan Council of Elders of Guatemala. They, of this Council of Elders, then elect one leader ... and that is who Don Alejandro is. He is a thirteenth-generation Mayan shaman—and he knows more about the Mayan prophecies and calendar probably than anybody alive.

When I met with him in Sedona, there were basically two things that he wanted. According to the Mayan calendar, there had to be exactly sixty people to represent the world, and so he asked me to get these people together to **be the world,** since the calendar says the world will be watching when the Mayan Nation begins to reconstruct the knowledge and the memory of their experiences of everything that had happened to them over the last twenty-six thousand years. So, I did; I showed up in Guatemala City with these sixty people from five continents and from twenty-three nations. We began this incredible journey, through constant ceremony (at different places and in different ways), to fulfill the Mayan prophecy.

While we were there, doing what he asked us to do, he began to tell us what the Mayan prophecy is ... but only parts of it. Only parts of it have been released. The really important part he hasn't released yet, but he has talked to me about it. Most of the world has only heard the interpretation of the prophecy that everything ends in 2012—the doomsday prophecy. He goes along with that, in a certain sense, since he specifically says that a physical pole shift is going to take place and that it will take place sometime between any minute now and the next five years. The axis of the Earth will rotate to another location.

■ *And what does he say, according to the Mayan prophecy, will happen when this occurs?*

You have to understand that it is all about cycles. When we look at the Mayan calendar, which is a wheel, and think of time ... (like we look at our watch and see it is five o'clock), we forget that this is representative of celestial cycles that are going on, like the rotation of the Earth. There are all kinds of cycles most people are not aware of. One of these cycles is called the Precession of the Equinoxes, and that is a little less than a twenty-six-thousand-year cycle, which happens to be ending precisely on December 21, 2012. That's a matter of science.

There is another cycle the Mayans are involved with and that is fifty-two hundred years long—if you multiply that by five you get twenty-six thousand. That cycle is also ending on the date of December 21, 2012.

■ *And what does that represent, the number five, in your calculation?*

It's a pentagon, of course, in geometry. It only links to this other cycle once in the entire time—it doesn't touch the cycle except once, every twenty-six thousand years: 25,771.5 years, to be exact. So, these cycles are tied not only to time and everything we think of in the classical sense, but they are also tied to consciousness itself. You might say it is the "cosmic DNA" (this is how the Mayans look at it)—an unfolding of events taking place in all the cycles, not only on the Earth but all the galaxies, everywhere, just as in human life there are certain cycles that, when they merge, make us change. We become something different.

They do say there is going to be a pole shift, which will mean massive destruction all over the world: this is when continents rise and fall, mountain ranges are formed . . . it is huge change on the Earth. At the same time that this cycle is ending, a New Cycle begins and that cycle applies, according to the Mayans, primarily to human consciousness.

■ *Well, a lot of people are very fearful of the idea of a pole shift and the dramatic consequences it would bring about in the planet. What do you tell people to help them understand this process, so that they move into trust and anticipation, not fear?*

The Mayans want to speak globally to the earth population, and that is what they want to speak of: how to **be** during these changes. They say there is a certain way, a certain vibration, that human beings can hold, which will put them in direct connection to the heart of the Earth and the sun. When you do that, there is no way you will be harmed; rather, your conscious perception of the world is going to be completely changed. You are going to perceive this one reality in a whole new way; it won't even be the same visually. This is the message they want to get out to the world.

It is not difficult—anybody can do it. You have to just know how to be. The Mayans talk about a "period of darkness," and they say the last time this happened was thirteen thousand years ago (the other end of the Precession of the Equinoxes). At that time, there was a period of darkness, and millions of people died... simply because they didn't know how to be. Had they known how to be, they wouldn't have died. This is what they want to tell the world: what people can do, so that they won't have to worry about the changes that are coming. That will be presented, probably before the end of 2010—maybe sooner.

■ *How will that be presented? Will those sixty people that you brought into the picture be spokespeople for this group of Elders?*

No, it will be the Elders themselves. There are other groups of Mayan Elders I work with, but this particular one is the most powerful. They are the ones who say they want to tell us about this, but they're not ready yet. They say they are "waiting."

■ *Why are they waiting ... considering what you said about it happening, possibly, any minute now?*

I don't know. I can't tell you that. I don't know why they're waiting—but they are....

■ *Perhaps they are working with earth energies, waiting to see how people progress in the interim?*

Every move they make is based on the Mayan calendar. They can see when they are supposed to do something and when they are not, and so, watching how they work, I don't ask too many questions on that level. I just know that they will do it when they are ready... and that, until then, they won't. But I do know that they are preparing to do this, because they told me so.

I've set up a worldwide global TV system, for over forty countries, that will be ready anytime this month, so that when they are ready, they will be able to go into a certain studio and talk to people all over the world.

■ *How can people tune in to that—through your Web site?*

Yes, my Web site will talk about this when it happens.

I would like to talk about what they say about the New Cycle. The whole world is focused on the end of the old cycle. What the Mayans are saying is that financial systems, governments, are going to die away.

■ *Well, yes, we're already seeing that happening now, before our eyes.*

Yes, and they're talking about a new way of proceeding. We begin to change in ways that, right now, we can hardly imagine. For one thing, they say that we will actually be able to communicate with the rest of the Universe, and all the light forms that are here at this time, as well. More will come here. We will actually have an open exchange with extraterrestrials, for the first time in a very long time.

■ *I inevitably lead in to this discussion, as I feel this reality is truly immi-nent. To understand that we are not orphans in this great Universe is so much a part of what I perceive of this galactic shift in consciousness ... when we get past this big "secret" of ET presence on Earth, and open commu-nication on a global scale.*

Yes, I agree. We have this idea that there's a war going on in the galaxy and there really isn't. Our galaxy is billions of years old and it has formed itself into a body that is very similar to our human body. There are regions that are similar to the liver, the heart, and the brain, and the rest, and they all work together. It is all based on love—the whole arrangement. Of course, there are tiny little places that are evolving that are causing problems, because they are children who are evolving ... and they don't understand the bigger picture.

The truth is: Life is here. At the moment, there are two hundred and fifty thousand cultures that are physically here, on Earth, that are watching and waiting for us to make this evolutionary jump. This is going to happen very soon. As soon as we move into the next level of consciousness, where we can perceive things differently, we will see them.

Life is going to be very different.

I was just in Colombia and I was living with the Kogi and the Arhuaco, which is another tribe connected to the Mayans. They're older—they're called the "elder brothers" by the Mayans. When I was talking to them, they talked about nine worlds and that we're heading into the "fifth world." This is what the Hopi and other Native Americans in the United States believe: that we're going from the fourth to the fifth world.

So I said, "That means we are going to the fifth world now—hopefully we are eventually going to make it to the ninth world."

They looked at me and said: "Drunvalo, you don't understand. Everything is about balance. There are four worlds on either side and the fifth world is in the middle. It is the most balanced of all worlds—so it is the highest world ... and that is what we are about to enter into."

This is a huge blessing for humanity, and there's no reason for fear at all. We're going to ride this through and it's going to be amazing. Because there are a lot of things I know are going to happen ... I know we don't have to be afraid. We may see destruction around us—that is Mother Earth realigning and moving her body around.

■ *Well, yes, but let's not forget a lot of it is human-made. Let's not ignore the fact that it seems we just cannot seem to get through the day, without creating war somewhere on the planet. As Stephen Bassett said: "War is what we humans do."*

I say: "We have to stop doing it!"

Yes, and that is the old cycle which is going to die and go away....

■ *So many people are so fearful of this change. My comment to them is: "What is it that we, the human race, have done of late that you are not willing to let go of? Take a look at what humanity has created in the last few years and, quite honestly, ask yourself: what are you clinging to?"*

Let's look ahead, at what we are capable of creating, rather than looking back, at what we are afraid to let go of.

Let's let it go, already!

We're very much like a butterfly right now. We're still in the cocoon, about ready to watch everything that we know disappear and something brand new appear. It's huge—it's beautiful! The Mayans are excited to tell us about all of this.

■ *I can't wait for the moment when they finally come out with their secrets and communicate them with all of humankind.*

I tell you one thing, with regard to what they say about December 21, 2012: they say that, almost certainly, it is not going to happen on that precise date.

■ *Yes, I agree. Let's not fixate on the date, which will come and go. Some will breathe a sigh of relief; others will experience disappointment—so I think it is important to move past that finite linear thinking and to understand that we are already "in transition."*

The Mayans actually think it is going to happen before that date.

■ *Drunvalo, when you say "it," what are you specifically referring to?*

There's way too much information than I can possibly share here ... but the beginning of this starts with the pole shift, according to the Mayans. A pole shift is always preceded by a magnetic pole shift and by magnetic disturbances that take place. We know this very clearly from the last time, thirteen thousand years ago, at the other end of the pole shift. Exactly where we are now, precisely, twenty-six thousand years ago, there was yet another pole shift.

■ *And we are, in fact, seeing a huge change in the magnetosphere now.*

Yes, that's right! The magnetic poles shift before the physical pole shift. Our magnetics have actually been dropping for about two thousand years. Five hundred years ago, they began to drop dramatically. In fact, about forty to forty-five years ago, the magnetic anomalies became so great that they had to change all the aeronautic maps of the world. That was done in the late '90s. And now, before they land a big jetliner, they have to check every single second as to where the magnetic field is, so that they don't end up landing out in a field somewhere.

■ *Hold it! That is some serious astrophysics that I'm going to have to ask you to elaborate on—especially since I will be on a plane tomorrow!*

The magnetic poles are changing every minute. They are not static. They are not steady anymore and haven't been for a long time. This is why the whales have been beaching themselves: they follow the magnetic lines, and now the magnetic streams have turned and gone into the land. And the birds that migrate around the world are suddenly ending up in places that no one has ever seen them before. They're doing the same thing—they're following the magnetic lines, but these magnetic lines are not the same as they were a hundred years ago.

■ *I think the whales are beaching themselves for other reasons as well—escaping from the unbearable sonic weapons being blasted throughout the oceans around the globe.*

As to air traffic: are you suggesting that a pilot, flying a jet, has to be concerned and aware of these magnetic anomalies?

Well, no, he doesn't have to think about it; in the control room they have an automatic program that constantly watches the magnetic lines. They can change in seconds! So the program has to constantly observe this and adjust to facilitate flights' landings.

■ *This is so significant for people who think of the Earth as a somehow fixed, guaranteed reality. They don't realize that, in a nanosecond, the magnetic fields of the Earth can completely shift.*

Nova did a show in 2003 with geophysicists who were studying this phenomenon, and they said that, at that time, the magnetic anomalies on Earth were so great that they believed there would be a magnetic pole shift in the next twenty-five years, causing a complete reversal, where north would become south and vice versa.

A year and a half ago, these same scientists went on the Internet to say that this was getting a lot worse, and happening a lot sooner than they had foreseen earlier. The government shut them down—they weren't allowed to talk about it. Only about nine months ago, they came on and talked for five days, before they got shut down again. They said that it was way, way worse than what they had predicted, and that this pole reversal could happen at any minute.

This is exactly what the Mayans are saying right now—that the pole shift could happen at any minute and that, when it does happen, it will be completely finished within twenty hours. You wake up one morning and the Earth shakes like you can't believe. It stops twenty hours later, and you are now in a completely different location on Earth, relative to the equator.

■ *I think it is so amazing that so many of us have truly come in to this incarnation to be part of this shift. There are so many old Atlanteans and Lemurians around ... old souls who came back to experience this moment in Earth's evolution.*

And remember—this is not governments orchestrating this. This is God—it is nature. You gotta go with it! It is just the way it is.

■ *If anything, the governments are generating the fear through so many control mechanisms that have been put in place on the planet. We have to get past that fear of destruction and change and think: "What will it be like? How will it truly be, to be part of such a galactic event?"*

The whole world has been looking at the Mayan calendar and telling us what it means! Don Alejandro says that everything that the world believes about the Mayans—everything—the Mayans themselves didn't say! We have heard about the Mayan beliefs from archeologists, professors, governments, religions, but he says that not one word, in over five hundred years, has come directly from the Maya.

He says these people who interpret Mayan don't understand what they are talking about. That is why they will be coming forward, to speak to us directly.

■ *I am sure it is going to be a great honor for you to be able to provide that global access for them, when they are ready to address humankind.*

How does this information about the magnetosphere and the shifting of cycles tie in with the Serpent of Light?

It's all related. The kundalini of the Earth seems like a different subject, but it's all tied in to the same cycles that the pole shift and the other aspects are connected to. The human body is a replica of the Earth. A human being can have a kundalini experience, where the energy that is in the base of the spine starts to rise up—almost like the sexual orgasm—but it is coming up a different channel. It's the same energy—but coming up a different channel. When that comes up, it has a different purpose than the other channel, which is to reproduce. It activates a person's desire to find God, in simple terms.

Once you have a kundalini experience, you really don't have another choice! You have to keep searching.

The Earth also has a kundalini; it comes from the very center of the Earth, which is like our base chakra, and it moves to the surface of the Earth. It is tied to these cycles we've been talking about. Thirteen thousand years ago, it moved into western Tibet, way up high in the Himalayas. A huge white pyramid was built over the top of where that is, even though there was never, ever humankind living there. The kundalini that comes out of the Earth has a very wide circle, reaching Tibet, Nepal, India, and China, and it causes people who live there to experience their kundalini rising. They become teachers.

■ *This pyramid, which you speak about—is it an actual physical pyramid? I've been to Tibet twice, years ago, and never heard of it, nor saw reference to it.*

There have been two recorded expeditions to it. One made it right to the edge and looked down at it, but couldn't reach it ... because they ran out of food. The second one actually got into it, took pictures inside it, and recorded it for Earth, so it is now fact.

■ *Where can one see these photographs?*

I don't know. They came to me and showed me those photographs. There was no writing on the walls, and they said it looked like it was built yesterday. They said there was an open door, an entryway, and there was no writing or images anywhere, except in the middle. There was another huge pyramid, and on the northern face of that pyramid wall, there was only one thing: that was the Flower of Life, which is the symbol of nineteen circles that I have been working with, which came out of Egypt.

At that time, I was teaching that, and they came to my house to ask me what the image meant, because they were trying to figure out who could have built that pyramid.

That kundalini, linked there, has now moved to the northern mountains of the Andes, in Chile. And that is what the book is about—the movement of that energy, and the thousands of indigenous people who are working with that energy, doing ceremony, which is part of this movement. There are one hundred twelve tribes right now, sitting in a circle there, because they know how important that energy is. That is the awakening—because what we are about to find is that there are going to be teachers to be coming out of South America— just like before, from Tibet, like the Dalai Lama and other Lamas coming out of Tibet. Now that light has moved to Chile. In fact, the Dalai Lama has just asked permission of the president of Chile for him and about ten thousand other Lamas to move there, because he knows that is where the energy is now. That's where the consciousness is.

Every time the kundalini moves, a brand new spiritual awareness is taught. We tend to think of spiritual traditions as somehow being fixed in stone, and that they are somehow classical and unchanging ... but they do change. They change according to the cycles of time, just like our unfolding DNA.

We're going to hear concepts and ideas coming out of South America that we have never thought of or contemplated before. It will also be coming out of the Pacific, because it is aimed over the Pacific. The original Lemurians, of the Polynesian world, from New Zealand all the way to Easter Island, will bring forward teachers. It hasn't happened yet, but it will happen very soon.

We live in amazing times—just amazing! Just to be alive on Earth right now is such a blessing. It is incredible that we have been chosen to be here, during this extraordinary time of change. Again, this is nothing to be afraid of. It is an honor to be alive here and to be able to experience the kinds of changes that are taking place now, because these enormous shifts don't happen very often. This one, in particular, hardly ever happens. This is what the Mayans are telling me. This one's different than thirteen thousand years ago, even twenty-six thousand years ago. This one is connected to an even longer cycle: the longest cycle in existence. It is coming to an end at the same time.

There are going to be changes beyond what the Maya and the Kogis and the Aboriginals of Australia and the Zulus of Africa are talking about. The changes are going to be even greater than what they believe is coming, following right behind the shift.

■ ... As we move into a higher dimension, where time no longer exists.

In the Mayan prophecy, they talk about a period of time at the end of these cycles called the "end of time." It is a window of time around December 21, 2012, which extends about seven or eight years around that time. That end of time started with a cosmic event that occurred just recently.

On October 24, 2007, a comet came into the solar system, which was given the name "Comet Holmes." That comet was no big deal to astronomers: just another comet that came into our system. But when it got near the sun, it blew up—it exploded—until it became a sphere of light. It kept expanding and expanding, until it became this huge mass of blue light—a perfect sphere of blue light! It actually became larger than the sun.

It is now recorded as the largest object ever recorded in history. That was prophesied by the Hopi, over two thousand years ago. It is now known as the "Blue Star" prophecy, and it has been recorded by the Hopi, and accepted by tribes all over the world, as the "beginning" of the end of time. At this moment, the entire world is now inside of this window called the "end of time," and this is where all these changes are going to happen. They don't all happen on that day (December 21, 2012); they could happen all the way to around 2015.

■ *So, as I said, we are looking at the end of time* as we know it.

So much is happening around the world associated with this. The change has been going on since it started about a hundred years ago—it didn't just turn on. We are no longer who we were a hundred years ago. Think back—the idea of going to the moon and into space were just unheard of, and now we're not only ETs ourselves, flying around in space, but we are changing in ways of understanding and knowledge, coming to us from all over the Universe. It is a time of incredible expansion.

The Mayans really do understand this, though I will say that they have said they lost a great deal of their knowledge and information, because of the invasion of the conquistadors. When they came in, over five hundred years ago, they destroyed almost ninety percent of all the Mayans' writings and records, so what they are doing now is reconstructing their knowledge, in two ways. One of them has to do with the brain. They have elected twenty-five of the Mayan people to rewrite everything—not the way that we, the technological man, has thought of it, but in the way they think of it. They are also doing it in another way, one that is hard for us to understand. The second way has to do with the crystal skulls. There are thirteen crystal skulls that the Mayans have created over the last thirteen thousand years since the last pole shift, when they were, at that time, in Atlantis.

■ *Hmm. There are so many different opinions about the source of the crystal skulls and their link to Atlantis. Personally, I believe the skulls in question are a lot older than that and that they are of an extra-dimensional source. I wonder. . . . Many of the Mayan shamans I have met, especially in Peru, feel more connected to Lemuria than Atlantis.*

The Mayans have told me directly that they came from Atlantis. They told me that their ancestors created these skulls once every thousand years, and that each of these skulls holds the knowledge of one thousand years. They stored it inside the crystal skull, very much like we do with our computers.

Computers can't work without silicate, which is the base of crystals. Without crystals, computers would not work. Crystal skulls, of course, are made of crystal—silicate—and they hold memory in the same way our computers do. This was discovered at Bell Labs, so it's scientifically discovered that crystals can do this. It is not outside of science.

All this memory of the last thirteen thousand years is now coming into the Mayan Nation, in a download from these crystal skulls (through very specific ceremonies) into thirteen modern-day Mayan shamans. That information will then be blended with the intellectual information of these other twenty-five Mayan people, who will then reconstruct the knowledge that the Mayans lost to the conquistadors. This is what they are doing right now, and this is what we are waiting for. When they have done that, then we will be able to hear information that has long been lost.

I can hardly wait.

Aho, Great Spirit! Neither can I, dear Drunvalo; neither can the world. We have so many different story-lines and so many differing experiences of what has come before . . . even of what is happening today! We can only imagine what will emerge as infinite truth, at the end of time, when we stand, hand in hand, with who knows how many beings.

Will we join together with beings from so many worlds beyond our own? Will we meet ourselves, in our past and future incarnations? In the no-time of our heightened awareness, fully aware of the Greater Universe of conscious life, everything is possible and, as you say, nothing is written in stone. We are creating it all, moment-by-moment, step-by-step, as we move through the Cosmos of Soul.

The New Dawn is upon us.

We are emerging from the chrysalis . . . preparing to take flight, beyond the matrix and into the light.

Read more about Drunvalo's works and future projects at:
www.onelotus.net

His books are:
The Ancient Secret of the Flower of Life—Volumes I and II
Living in the Heart
Serpent of Light

The Secrets
of the Crystal Skulls

■ ■ ■ ■ ■

Stephen Mehler
(Shown above with Khemitian Wisdom Keeper Hakim Awyan)

Stephen Mehler, an archeologist, historian, and metaphysician, is one of my greatest discoveries: a true "soul" connection. As the torchbearer of the Khemitian wisdom, he carries a wealth of experience and knowledge that he acquired through his long association with my beloved Hakim, the wisdom keeper of the Khemitian tradition of prehistoric Egypt. Hakim passed into the light August 23rd, 2008, but a number of us, like Stephen, are determined to forever bring his light through to the world. That knowledge is encapsulated in Stephen's earlier books, *From Light into Darkness and The Land of Osiris*. It is also discussed in my book *Where Pharaohs Dwell*. On this occasion, however, we will be exploring another aspect of Stephen's research and experience, which delves into his new book, *The Crystal Skulls: Astonishing Portals to Man's Past.* It is co-authored with David Hatcher Childress, with whom many of you will probably be familiar. I've asked him to blow the lid off on crystal skulls, with a particular eye for truth and grounded information, as well as the metaphysical interpretations of his experience with some of the truly ancient crystal skulls. I certainly have my own opinions and powerful experiences about the crystal skull phenomenon and its links to Tibetan and Mayan traditions—to Atlantis and to the Sirian star system. Stephen and I don't agree about everything and that makes our dialogue all the more exciting, as we explore our shared knowledge and our diverse areas of expertise. He is a man I deeply respect and an ancient soul brother I have only just recently come to rediscover. This is the magic of life—this is the power of the crystal skull connection.

■ ■ ■ ■ ■

■ *Welcome, Stephen. You know how I feel about you and the wisdom you carry about so many subjects. It just amazes me that they are so utterly synchronistic with my own journey in this lifetime.*

Thanks, Patricia. I do so appreciate our connection, because, as we have found out over these past months, we have so many deep connections, on so many deep levels. It started with the fact of Hakim's passing bringing us together, and then finding out that we lived in the same towns around the world at the same times ... and this deep connection with crystal skulls, which is not unrelated to our fascination with ancient Egypt, is magical.

■ *In fact, if you really look at it, I think we both can agree that we have a "Sirius" connection—pun intended! It's so delightful because, of course, my work also involves channeling the Sirian Elders and I have the great joy and honor of connecting with many people who are attracted to my work, who are linked to Sirius—as I'm sure you are. There is, of course, the unfolding story of Sirian influence in human development that gives us this incredible sense of having that Sirian star system wired into our DNA!*

So it seems to be all related together: Egypt, the skulls, the Mayans, the Khemitians ... Atlantis. In fact, I look at your own programs and schedules and find we seem to be always headed toward the same places at the same times. It's total synchronicity.

Absolutely. And what makes this interview fresher and more vital is that I now have read your first two channeled books from the Sirian masters, and so I now know what you channel about skulls and about Sirius. And since you have read my book on the crystal skulls, we can get to talk about that in much greater detail, now that we both know what the other has written about the skull phenomenon.

And while of course a lot of the material about crystal skulls is **speculative**, we can talk facts. We can talk about the history and the actual data and other evidence that we have. Yet, when it comes down to crystal skulls, so much of it becomes subjective, because as **you** have channeled about them, so many others have as well. Many people have channeled with them, including me; it seems like everybody who gets around a crystal skull starts channeling.

We can separate some of the information, but it isn't that there is no real truth here. I mean, we can put down the evidence. I've been involved with them for thirty years now. But it still comes down to opinions and ideas. What will impact people will be how they resonate to what we say. So, you know, we can say this and we can say that, but a lot of it is so much speculation. A lot of it is so much right-brain as opposed to left-brain, so the best approach is to present our perspectives on it, and let people decide for themselves.

■ *That's right, but what's exciting and important is to offer a balance between the two perspectives. I am a right-brain-dominant person all the way, but I do try to honor the left-brain people and merge these two aspects of the human mind together, so that we can bring it together into one beautifully balanced male-female mind. And the skull really does that, doesn't it?*

Oh, yes. They were absolutely essential tools for the ancient peoples, whoever created them. We can speculate about "who," but what we know is that crystals are definitely devices for connection for interdimensional energies.

■ *Indeed. They may very well have deliberately brought us together! You know, it's just amazing. You mentioned our strong link at the beginning of the conversation, and to think: you were directing the Rosicrucian Museum headquarters about three blocks away from where I lived, back in the 1980s!*

Well, I got involved in crystal skulls first in 1979 or 1980 . . . and there you were, living just a few blocks from Rosicrucian Park, where I first discovered them, so yes, it is pretty amazing synchronicity.

■ *It is . . . and that's not the only amazing thing. Probably that same year that you were getting your hands on them, I laid eyes on my first crystal skull (in this lifetime, at least!). I found the most outrageously beautiful smoky quartz, human-size skull, with a rainbow fracture right through its third eye, in an obscure little crystal shop in the Bay Area. It is where I first discovered moldavite, over twenty-three years ago. At the time, the skull was priced at five hundred dollars; today it would probably sell for ten thousand dollars. But at the time, five hundred dollars was really big money for me—and it is still, for a crystal—but what a crystal!*
Unfortunately, I was with my mother, who was the more rational of

the two of us. I told her I just had to get this skull, and she reminded me of how more wisely I could spend that money! So, I left it, and yet it still talks to me.

That was the first really big moment of connection, which has taken me on an exquisite journey of exploration with the skulls.

You connected, for sure. This is a very typical story that we hear from people. We have always said that we don't go to crystal skulls; they call us. They seem to have this ability to know who are crystal people, who are skull people, who are "connected." This skull was obviously calling you. So it was almost like imperative that you had to get it. And it's been the same with me. They just pull you to them.

I've heard so many stories—from so many people—that are similar, that they go into shock (having no idea about crystals, no idea about skulls) when all of a sudden they see a skull . . . and boom! Their life is changed. They just know they have to have it. The skull calls to them and talks to them and tells them its name, amongst other experiences. As well you know, people develop a personal relationship with them. They're not just little objects out of inanimate, inorganic material that sit on the shelves as decorations. We actually develop intimate, personal relationships with them.

■ *When I finally got my first crystal skull, my partner, Franco, just said: "Why do we have to have this in the house?"*

He's Italian and these symbols, for most Italians, are very different, very frightening, and very dark, you know, because of the Catholic, religious influence here.

Exactly!

■ *"I just have to have this skull," I told him. Like you just said, I couldn't explain it. I just had to have it. Now, that lone skull has developed into a*

Committee of Thirteen Skulls. One, which you know of, is most likely a very, very ancient Tibetan skull.

Yes, it's a very interesting skull. And I was very glad you sent me photographs of it. There have been rumors, for years, about where they came from; for example, Atlantis—that is a predominant one. There are also stories that there were thirteen skulls that came from Tibet: thirteen very ancient skulls. For years, people have been talking about this skull, maybe from Atlantis . . . that skull, maybe from Tibet. But there has never been anything definitive about them.

However, yours definitely looks like one that could be an ancient skull from Tibet. And we know that the ancient Tibetans were very interested in crystal and probably did a lot of crystal carving.

■ *I went twice, in one year, to Tibet, back in 1997. It was before the Chinese government really moved in all the way, definitively breaking the energy there. I found crystals in the weirdest places. You can buy crystals in crystal shops in our Western realms, and often they are presented as "Tibetan crystals," but I know for certain that the crystals I bought there were Tibetan, because I bought them there—in the markets. They have such an incredible energy, very different than a lot of other quartz that you work with. Crystal workers, like you and me can, sense a very high vibration.*

It also depends on how they were worked. We always talk with people about how we can tell an ancient one from a modern one, from contemporary ones. More modern skulls, of course, are carved to be aesthetically pleasing, using the finest quartz or minerals possible—made to look very much like a human skull. But the ancient skulls sometimes actually look ugly or crude, because that was the intent of the carver. We say this over and over and over again. It's the intent of the carver, and so many of the ancient ones were not carved for aesthetics. The crystal was chosen for its metaphysical properties, and the skulls were used as powerful tools: meditation tools, psychic devices, healing devices, etc.

Sometimes, when you come across one like yours (that may very well be very ancient), to be used in rituals, the energy involved with it is so different, that you can immediately tell it from a modern, contemporary skull.

The true ancient skulls have so much power; they have so much memory. There is so much human energy involved with them, because of how they have been directed. They've usually been used in rituals.

I've been taught by masters how to determine which ones are the real ancient ones: they're usually not aesthetically pleasing, and not of the finest-quality quartz. But the energy, the power, and the memory are incredible, because they've been used in rituals for so many thousands of years.

■ *In fact, this skull, Norbuk, has that intense level of energy. I bought it from an antique dealer in Vienna, and the story of how I found him is quite bizarre. Someone sent me a picture of this amazing skull, that had Tibetan ornamentation around it. Really, really, I've never seen anything like it before or since. This really incredible ornamental design around it was Tibetan, similar to some of the more elaborate Tibetan jewelry, which had coral and turquoise and Tibetan silver. It was showing in a museum, somewhere in Austria. It even had the separate jaw, like the Mitchell-Hedges Skull.*

He investigated and found that the person who had supplied this magnificent skull to the show was an Austrian antique dealer, so he went to the shop to get information about it. He was told that the skull was not for sale and that it had gone back to its owner in Dubai, but then asked: why the interest in it?

This friend told him that I was eager to find it ... and why. After a short time, the dealer came out with these other two skulls. One of them was the one that I eventually purchased, and a second, other one, much smaller, went to this man who made the contact for me. The dealer, who was a reputable dealer with excellent credentials, told him that these were from the Himalayas and that they were found in a Tibetan temple.

Wow.

■ *What's interesting about the skull is that it had a patina on it, which I partially removed to be able to scry with the skull. I haven't completely cleaned it, as I feel this patina holds a lot of history—in fact, I would like to see if I could get this material examined. I'm quite sure it's going to have something like incense or yak butter in it, and that there may be clues, from that substance, as to its age and original location.*

That's right. Actually, your skull looks like it had been buried, because there's a lot of dirt and grit residue. It might actually have been buried in the temple for a while.

■ *Indeed. It might have been buried to protect it.*

The bringing up of your skull, and how you got it in Vienna, leads to what we can discuss about the known history that we know about the subject. Crystal skulls really exploded on the European scene and that's why it's very possible to find some real ancient skulls or some interesting skulls all through the region—particularly Austria and Germany. Crystal skulls really exploded on the European scene in the middle of the nineteenth century, when the Austrian emperor Maximilian conquered Mexico. This was the 1840s to the 1850s, and so then the Germans came into Europe, and particularly to Brazil, and started mining quartz and obtaining crystals.

There was a great interest in crystal skulls in Europe, in the late nineteenth century. This is really why there are probably a lot of them in collections in Europe. And that's why it's a good place to locate them. They really started coming out in Mexico, in the tombs: Mayan tombs, Aztec tombs, and Olmec tombs. That is when they really exploded on the European scene, but in the West, nobody knew anything about them.

■ *In fact, I would say as early as ten or fifteen years ago, almost nobody even talked about skulls except a few of us—like you and me.*

I would say even earlier, when the gentleman I'm going to talk about, Nick Nocerino, came on the scene. Before the '50s, there were very few people who knew about skulls and, not to mention, the Mitchell-Hedges Skull came into play. There were a lot of adventurers like Mitchell-Hedges, who had gone to Central America and Mexico, and they started to talk about how the shamans had crystal skulls. That's when it really exploded on the scene. But I would say before the nineteenth century, there were very few people in Europe and in the West who knew anything about that. The Mayans did, of course, as you well know.

■ *Why do you think it particularly exploded in Europe in the 1900s?*

They were discovering South America, Central America. Germans had always been some of the leading archeologists in European history, along with the Italians. Germans started going into Brazil and into South America. Again, Maximilian, the Austrian emperor, had Mexico ... and so they were looting tombs. They started to loot tombs throughout Central America and Mexico and, consequently, they discovered crystal skulls. They started to bring them back to Europe and then they actually started carving modern ones in Europe.

People realized that Europeans might find them interesting—that they might want to have some in their personal collections. So, they started carving crystal skulls. We don't know for sure about the carving of crystal skulls. It's very interesting and we can talk about this if you want to in detail, with respect to the Mitchell-Hedges Skull.

There was one particular city in Germany; it's a twin city now, called Idar-Oberstein. The city of Oberstein has been the leading carving area of quartz in European history. It has not been definitely proven yet, but there is no doubt that the German and Austrian "soldiers of fortune" went to Mexico and Central America around the 1860s to the 1880s, and then these crystal skulls started to appear in Europe.

■ *This is what makes mine so interesting, because, like you said, it's not a particularly attractive skull. This dealer had these skulls packed away and it was only because this friend of mine insisted, talking about the more spiritual aspect of the skulls, that the dealer decided to show them. Apparently, he brought them out and he had never taken them out to show—yet he felt he had to show them to this fellow.*

It's probably the skulls themselves that told him that. As you said, it usually works that way.

Yes, and again it's when I look at the skull . . . I get this feeling that it is ancient. Now again, you know, to be honest, I always tell people that to make a really strong determination, of course, you have to experience the skull: you have to see it; you have to observe it. You have to look at it like a scientist, with a magnifying glass, and then meditate with it, to really get the feel of it. But when I look at pictures, I can get a good sense right away, because of my experience with them. When I see some of them crude like yours, some of them untarnished like that, and other aspects like that one exhibits, I tend to think it is definitely not modern.

Because modern skulls (particularly those that have been carved in the last twenty to fifty years) are carved to be sold, they're going to be cut from a good piece of quartz: very smooth and at times anatomically correct, like the human skull. The great anomaly is the Mitchell-Hedges Skull. That is the one that is the most confusing for people. I think it's the most misidentified because it's so exquisite— it's so beautiful. And yet, you know, there are those who have channeled that it is one of the original thirteen skulls of Atlantis.

I reject that. I originally thought it was ancient myself, when I first looked at it, because it's very powerful, very, very powerful.

■ *It sure is! It is immensely powerful. I have meditated with it. I think you have too.*

Many times.

■ *Of course, we're going to both agree on one very central thought ... and that is that if enough people meditate with a piece of quartz and empower it with that kind of consciousness, in essence it also resonates to that much love and adoration and intellect.*

Without a doubt. That's really one thing we should establish right from the get-go, so that people don't misunderstand when I get into that part of the discussion about the ancient skulls. **All quartz is ancient.** All quartz is millions of years old, unless it's synthetically produced in the laboratory. All quartz is ancient; all quartz has energy; all quartz can be programmed. All quartz can be used for healing, for storing of information, and for communication on interdimensional and extraterrestrial levels. And then again, you can even take a modern skull, put it with an ancient skull: the ancient one will download to the newer one. It's like I often say to people: the ancient skulls can be thought of as "mainframe computers." They have all the hard drive, all the memory. You can take any quartz skull (that has been cleared of all memory), even if it has been carved last week, and then you put it with a Max or Sha Na Ra—or maybe with **your** skull—and immediately the older one will download to the younger.

■ *Right. And can I just interject something here? You and I both also knew Marcel Vogel, which is another incredible connection. Marcel, for the people that don't know who he was, was the developer of liquid crystal.*

Actually he got his PhD in liquid crystal. He was a PhD in chemistry.

■ *... and he contributed this knowledge to IBM. That's how he made his money... before he dedicated himself to healing.*

Yes, he was the first great one that I worked with. I worked with two giants in the field who passed away: Dr. Marcel Vogel and Mr. F. R. "Nick" Nocerino.

■ *You actually worked with Vogel? I didn't know that you worked with Vogel with crystals—that is real credential-building!*

Oh, sure. Vogel was the first. He came into the Rosicrucian lab when I worked with the first crystal skull, close to twenty-nine years ago. Marcel Vogel lived just a couple of blocks from Rosicrucian Park. I worked with two crystal skulls with him: one we called the "Mayan Crystal Skull"; the other, the "Amethyst Crystal Skull." I actually have both sessions recorded on audiotape.

Vogel was the first great scientist who I worked with. He was able to walk in both worlds, as I do. He was, as I mentioned, a PhD in chemistry, and he worked for IBM as a senior scientist, for over twenty-five years. He also worked with liquid fossil resins as crystal. He wrote many great papers on the subject. He produced scientific theories, which are in my book, which I used. I actually use his scientific theories that regard how crystals can store information. They are still viable today.

And he was also a Rosicrucian.

Then, the greatest source on the subject of crystal skulls was Mr. Nick Nocerino, who also was a great, one of the greatest metaphysical parapsychologists who ever lived. He knew more about crystal skulls than anybody who's ever lived.

■ *How unfortunate for me, I have to say, because I was right there in the San Francisco Bay Area. I had not one but two appointments with Nick and both of them fell through, for various reasons. I just missed having the chance to meet and to have worked with him.*

I have regrets, too, that I didn't spend more time with him. There were some different reasons (I don't want to get into personalities and conflicts and politics) that I didn't spend more time with him. I wish I could have, but I treasure the time I was with him and the information that he shared with me, which was superlative. And he worked of course with the Mayan shamans, having gone to Mexico.

■ *Let's talk more about Nick, because most people don't really know about him. Let's be honest: there are a few names that are out there, position- ing themselves right now as the "people" or the "voices" for the crystal skulls but, quite honestly, I think that we can't talk about anyone as a knowledgeable exponent of crystal skull information, without first talking about Nick Nocerino.*

In fact, one of my favorite Nick quotes was: "The more I study crys- tal skulls, the less I know about them."

He was immensely humble. There was nobody who knew more about them than he did.

It's a lifelong study. You never can come to the end of the information and the knowledge about crystal skulls—this you realize, the
more you work with them. It happens to be by default that I'm now
one of the senior researchers in the field. There's actually maybe
one other person, who was a student of Nick's, who, I would say,
had been involved with crystal skulls longer than I have. But since
they both have passed away and I first got involved in 1979 (the first
time I saw my first crystal skull was January 4, 1980), twenty-nine
years ago, I guess (by nature of the fact that I am still around) I am
one of the senior researchers in the field.

■ *That's amazing. That was the time that I was walking away from the*
beautiful smoky quartz crystal. That was the time.

So I imagine that is actually when I first "met" you.

■ *. . . through the crystal skulls!*

Well, Nick Nocerino was the first person in the West to give a public
lecture about crystal skulls—in March of 1980. Nick was an Italian-
American, who was raised in a typical (you **think** typical) Italian
Catholic family. However, the difference with Nick and why he was
who he was is that his maternal grandmother was a Wiccan High
Priestess. There was a tradition of Wicca in his family. She was called,
in Italian, a *strega,* or *stregona,* which means High Priestess. He also
had a woman named Comadre, who was his teacher as a small boy.
In fact, the story even starts with his birth. Nick was born on the
kitchen table, in his family home in Brooklyn, and his grandmother
and her coven served as the midwives at his birth. So, clearly, immediately from his entrance into the world, he wasn't brought in as a
typical Italian Catholic.

He was brought in under Wiccan influences.

■ *As I recall, in the book you talk about the fact that he had crystals around him at the birth table ... is that right?*

Yes, as he was born, according to Wiccan tradition, they did not discard the placenta. It was given to this lady who was known as "Comadre," a very close friend, a Wiccan High Priestess and friend of his grandmother. She took the birth placenta and put it in a mason jar, with a bunch of small quartz crystals, to be charged by his birth placenta energy. It was the connection from his mother's energy that they believed the ancestors' energy would all come through. When he was eight years old, he was told to go to this lady's house. He knew nothing, up until that point, except the fact that he was born very psychic, very sensitive. He just assumed everybody was. But he was extremely sensitive. He was already in communication with spirit, before he was eight years old.

When he went to her house, she gave him the crystals to wear around his neck and then discarded the placenta, so then he would wear the birth crystals around his neck as power, and as protection. They taught him about the use of crystals: how to use crystals for healing, what were crystals all about.

He started out from a young age and, by the time he was eight or nine years old, he already knew about crystals. He was nine years old when he had his first vision of a crystal skull in his home in Brooklyn. He was walking up the stairs to the second-floor landing, where there was a large mirror facing the stairs. As he came up the stairs, he looked into the mirror and he immediately saw a skull appear.

Having had the Wiccan background and his psychic experiences, he wasn't frightened. He looked into the mirror and he saw that the skull was not a human skull, but rather it was a crystal skull. He then saw two very significant things: he saw a cobra come out of one eye; he then saw a jaguar come out of the other eye. He never forgot that image. In fact, he had somebody draw it for him later. It was later that he found out, when he connected with the Maya, that it is the Clan of the Jaguar, which keeps the knowledge of the crystal skulls of the Maya.

■ *I am also of the Jaguar clan, and I am sure you are too.*

Not surprising! As it turns out, that image led Nick to go to Mexico in the late 1940s. He sought out and connected with the Jaguar clan, and then met the gentleman that we eventually got involved with, who brought the Mayan Crystal Skull and the Amethyst Crystal Skull into the United States. His name was Francisco Reyes. He claimed to be a Priest in both the Aztec and Jaguar traditions; he claimed that he was of the Jaguar clan.

I met Francisco in November of 1982, when he brought in the Amethyst Skull, and I met him again in the early spring of 1983, when we did a ritual—together with the Amethyst Skull.

■ *You've mentioned your own experiences with the Amethyst Skull and I would like you to please share with us what you experienced when you worked with it.*

I was working at the Rosicrucian Park, as a staff research scientist, in November 1979. We got a notice from a gentleman who was curator of the Rosicrucian Egyptian Museum. He had been contacted by a gentleman in San Jose named John Zamora, who said he had an object he would like us to look at, to see if we could authenticate, document, and maybe even exhibit in the Rosicrucian Egyptian Museum. Of course, as it turns out, this gentleman, John Zamora, was an agent for some Mayans in Mexico, who were trying to sell some black-market antiquities. They were trying to use me and the

Rosicrucian order as agents, to try to help them to do that. What he first brought to us was a series of photographs.

Again, this was **it** for me. As an archeologist, as a prehistorian, I have heard a lot about crystal skulls. I had seen pictures inside the Mitchell-Hedges Skull, but it had not really resonated with me at that point—not really sunk in. When I first looked at these photographs, in November 1979, of what we later called the "Mayan Crystal Skull" (named by Nick Nocerino), it made a really deep impression on me.

I said, "Wow! This is something else. You know, I would really like to work with this!"

We asked this gentleman if he could bring the skull in, for us to work with. That's when it started—January 4, 1980; that is when I first experienced an ancient crystal skull. The Mayan Crystal Skull was definitely ancient.

■ *What makes you say it was "definitely" ancient?*

... Because we worked on it: work that I did on it; work that Nick Nocerino did on it. Nick Nocerino, remember, first got involved with crystals at the age of nine; he saw his first crystal skull in the South of France in 1944; he went to Mexico in the late '40s and saw many ancient crystal skulls. By the time the '60s and '70s rolled around, he was the expert. He went with a group to Mexico, where he met Francisco Reyes in the Oaxaca area. (I mention that to you because you know Mexico and have been there.) The Amethyst Skull came out of Oaxaca.

Nick paid his expenses to come to the U.S. to bring the skull in. How did this gentleman, a very poor, not terribly well-heeled Mayan Priest, come in and out of Mexico and the U.S., carrying ancient crystal skulls through the borders, and never checked once at customs?

Amazingly, he came back and forth with all these crystal skulls. He came to the U.S. in October 1979 and worked with Nick's group, which Nick had founded in New York City in 1944.

They incorporated in California in 1945 and stayed together until Nick's passing in 2004, when the group disbanded.

The Society of Crystal Skulls International was the most reputable, the most scientific, the most long-lasting organization to study crystal skulls. So Nick founded that organization and developed all the students. He became the first person to get a credential to teach parapsychology in the state of California.

■ *He was amazing. But let's get back to the skull.*

To answer your question, I'm just trying to get to the point. Nick worked with the skull for four months. He was convinced it was ancient. I worked with it. I had a whole, full-day session with it, with Marcel Vogel. Marcel Vogel immediately pronounced it as "ancient," working with it. I had many people work on this skull, not in the presence of each other. We had questionnaires that people filled out. We recorded audio recordings of their sessions; I have hours and hours of audio recording of people working with crystal skulls.

■ *What kind of information came forward?*

Well, what was consistent was that people would consistently see Mayan scenes. Something like eighty percent of respondents would see Mayan scenes in it. That's why it was called the "Mayan Crystal Skull." It was a cloudy quartz skull about five inches high, and it weighed a good ten pounds; it had particular indentations on the sides, where our temporal lobes are, which was very particular. There are a couple of skulls like that, so it was obvious that they were all carved by the same group of people. The Amethyst Skull had these indentations too.

The piece was crude, although the carving was good: fine and smooth. And we saw so many scenes in it. It had a lot of high energy around it. So I came to the conclusion that this was an ancient artifact. I now know so much more, because that was the first one I worked with. I would still classify it as an ancient artifact, but not as old as others we could talk about.

■ *Well, let's talk about the others!*

Nick Nocerino and the Society of Crystal Skulls International came up with a scientific name to classify them. It broke them down into three levels. What we would call a "contemporary" skull is one that's been carved in the last fifty to one hundred years. The majority of crystal skulls you see out there, owned by people who claim that they are ancient, are actually contemporary skulls, carved somewhere between the last fifty to a hundred years.

Then, we classify what we consider an "old" skull as being more than one hundred years old. What we classify as "ancient" crystal skulls, however, are those that have actually been carved over a thousand years ago. I would classify the Mayan Crystal Skull as ancient.

Later on, after I left the Rosicrucian order, John contacted me again—in November 1982. He said the Priest had returned with another skull. I rushed to his house, and there was the Amethyst Skull. Bear in mind, I had not yet met Nick Nocerino. So I was not aware at the time that there were many amethyst skulls out there. I had never heard of that.

You know, he did have visions when he worked with the Mayan Skull of skulls with different colors. But I didn't consider skulls of different minerals as being "old." Now I can tell you that there have been crystal skulls carved in every possible stone you can think of: from amethyst to ruby.

■ *I even used to have a little small crystal skull, of course contemporary, carved out of moldavite.*

Moldavite? Wow! That would be very nice: that's **the** extraterrestrial stone.

■ *Yes, it was fabulous, while it was with me—but it leaped out of my hand one day and literally disappeared. It was a very small one, as you know how expensive moldavite is. But the power was immense.*

I was with the Amethyst Skull, and amethyst is my favorite stone. So I nearly freaked out with this one. I loved it. I met Francisco Reyes then and he told me so much about crystal skulls. He gave me much information. And the amazing thing, Patricia, is that I don't speak Spanish. Francisco didn't speak any English at all, so John Zamora served as an interpreter. Yet, there were times (you know the Mayans can do this) where he would look at me and he would communicate something in Spanish. I knew exactly what he was saying in my mind. He had that ability. He demonstrated that ability to be telepathic.

In fact, Marcel Vogel said that one of the major uses of the ancient crystal skulls (particularly the Mayan Skull) was that they are language decoders. In other words, you could have a crystal skull in front of you, have people from all different cultures, speak all different languages, and, through the use of the crystal skull, everybody would know what everybody else was saying.

This is almost like *Star Trek*, you know?

■ *For sure, it is uncharted waters for a lot of people. You know that I agree with you—even when we talk about the Atlantean crystal skulls, it's such a metaphor for a communication system. You consider the possibility of twelve crystal skulls being united, to activate a thirteenth, a sort of Master Skull or a hologram, which is what I present in my material. When they are all reunited, it activates. How perfectly this describes some kind of ancient computer network!*

Let's talk about the idea of thirteen original skulls. We all have visions. Nick Nocerino was the first one in contemporary society who mentioned seeing visions of twelve skulls, with a massive thirteenth in the middle. He mentioned that before anybody else was talking about them. When I discussed skulls with Francisco Reyes in 1983, he never mentioned any Mayan tradition of thirteen original skulls that had to come back together. That doesn't mean there wasn't such a prophecy: he just never discussed it. Not one of his contemporaries has discussed this.

You did, in your books, ten years ago.

People started to talk about this in the late '80s; they started talking about seeing visions of thirteen skulls. Then the idea that there were thirteen original skulls began to surface and to be channeled from different people. I wanted to ask you if you have learned more about this from the shamans, because I know you have now been initiated into the Mayan wisdom tradition in Mexico.

■ *Yes, and my initiation, quite appropriately, was consecrated in a crystal skull ceremony by the Mayan Keeper of Palenque.*

That's amazing. This was one of the reasons I wanted to go this year. I was going to do a tour, in connection with the Mayan shamans, because I have only spoken to two. I spoke to Francisco Reyes and a contemporary gentleman, who lives just north of Mount Shasta in

California, who is a Mayan day keeper. I worked with him with facts; we talked a lot about crystal skulls. He says that, as far as he knows, there was no Mayan legend tradition of thirteen crystal skulls that had to come back together. Again, it doesn't mean that there isn't one. I'm just saying that it suddenly has become very popular, in only the last ten or fifteen years.

■ *No, it doesn't mean that at all. Real shamans don't really say that much. They believe that silence holds the power....*

Exactly.

■ *My experience with the Mayan and Peruvian shamans, just like the Tibetan monks, is they say very little. It is the seeker who has to discover the wisdom, simply by* being in the energy. *Let me share a story:*

As you said, ten years ago I was writing about this Circle of Thirteen, although it was many more years prior that I was having visions of it and collecting crystal skulls of every kind of crystal and mineral.

I took a spiritual group to Palenque in 2005. Before setting off for this program, I told people that we would be doing a crystal skull ceremony at Palenque, and that we would be uniting twelve crystal skulls during that event. I kept hearing (from my Sirian guides) the message: "Call in the thirteenth.... Call in the thirteenth!" I heard this over and over again—for days on end. I interpreted this as meaning that I was supposed to bring twelve crystal skulls and that, somehow, a hologram or essence would appear in the middle of the circle. So, I started acquiring and assembling twelve crystal skulls to bring to Mexico.

The tour guide was concerned that I had no specific idea of how I wanted to run the ceremony. He wanted to establish a kind of ritual that adhered to his idea of traditional Mayan ceremony. I kept telling him not to worry— that I would know when it was the right time.

We got to the magnificent site of Palenque. It started to rain and he ran over to get shelter in one of the local stands selling handicrafts, where he ran into the father of the Keeper of Palenque, Kin. Kin asked the guide why he was there and they engaged in the usual small talk, in which the guide told the Mayan that he was guiding a tour "for this woman from Italy" who was planning a crystal skull ceremony at the site.

Kin asked him how many skulls were involved, and the guide responded that I had twelve skulls with me.

He said, "Wait here." Minutes later, the shaman, Kayum, appeared and asked him to reiterate the whole story.

Kayum asked him, "Who is this woman and where is she from?"

Later, I learned that he was trying to find out where I was from, because he had had a vision (he said it was prophecy) that a woman was going to come from "the north" with twelve crystal skulls—and that would return his power to him.

The guide kept telling him I was from Italy, which didn't fit the prophecy, so Kayum insisted, asking repeatedly if he was sure I was from Italy. Finally, the guide said, "No, in fact, she lives in Italy, but she is not from Italy. She's from California."

Kayum then said, simply, "I will conduct the ceremony. And I will bring the thirteenth skull."

Can you imagine that? He presented himself that night to my very fortunate group of people, with his majestic Mayan Skull. You can imagine the impact this had on us all.

Wow!

■ *And it's like, "Hey, everybody—have I got a surprise for you!" In comes Kayum with the Mayan Skull and of course this was the manifestation par excellence of "calling in the thirteenth"! We did the ceremony in Palenque with Kayum directing us and, remarkably, in the end he gifted me with another skull, "Estrella," which he said had been in the family for generations.*

He said that this, too, was ordained in prophecy and that I was to take it to Europe, to work with it here.

That just shows how connected you are. You see this is . . . what I'm saying is that this is **right.** When working with the Mayan Crystal Skull in 1980, I too saw a scene of twelve in a circle with a massive one in the center. In fact, when I saw it and when others saw as well, was that there were twelve different-colored minerals involved. They were green like jade. They were red like ruby. They were many colors.

■ *That validates! I have twelve different-colored skulls here in my temple.*

You see, this is real! I'm not sure about there being thirteen original skulls from Atlantis. There are people running around the world saying they have one of the original thirteen skulls. Nick Nocerino saw anywhere from fifteen to eighteen skulls that he would have classified as genuinely ancient. What we believe is there may have been many sets of thirteen. Now, whether people believe that or not doesn't matter. If you see these thirteen skulls as coming from Atlantis, that's fascinating; what I don't agree with is when people say there are **only** thirteen original ones. . . .

■ *"Only" is a word that has to disappear when you're talking in metaphysical constructs. Everything is possible.*

And again, we use the term "ancient" for crystal skulls that are over a thousand years old. That means a crystal skull could have been carved one thousand and one years ago and would still be classified as ancient. Yet, another one could have been carved fifteen thousand years ago, during the time of Atlantis, and that's also an ancient one. So there were sets of them, I'm convinced. Whether there were thirteen original skulls from Atlantis, or whether they were from Tibet, as it is said . . . I find in your book *Atlantis Rising*, if I just substitute for the word "Atlantis" for Khemit, I have no problem with anything that you say about them.

■ *Ha! You are such a Khemitian!*

Poppa [Hakim] would have been the same way. Everything that you say about Atlantis in the book—and there's some wonderful channeled material there that's from ten years ago—is unfolding. Here we are ten years later: certainly the warnings about the New World Order, about HAARP and global change and all the rest, is happening—particularly about diet. I found that very interesting, because I've been a vegetarian or a partial vegetarian since 1972. I haven't eaten red meat or chicken since. And here, you know, the Sirian Elders are telling us about eating at the lower end of the food chain.

When I substitute Khemit for Atlantis, I have no problem with any of the material presented in *Atlantis Rising*.

■ *Well, in my opinion, it really doesn't matter whether we call it "Atlantis" or "Khemit." Who knows? We could be talking of a civilization of over fifty thousand years ago. In your opinion, it does, of course, because you're an archeologist. But when we're talking about the ancient world I think it's all swirl: there are no boundaries; there are no fine lines to divide nations or countries, as we perceive them today. The borders that we identify now were most likely drawn very differently back then.*

The fact is that we were much more spiritually advanced over ten or fifteen thousand years ago than we are today. And that's what a lot of people are finding out. Now I would tell you the Mayan Skull, the Amethyst Skull, I classified them as ancient, but when I came across

Max and Sha Na Ra, they're much, much older. These guys are certainly two of the oldest I've ever come across. Both Sha Na Ra [Nick Nocerino's skull] and Max [the Texas skull] we can talk about. They're both definitely carved over ten thousand years ago. There's no way anybody in contemporary times carved Max or Sha Na Ra. I believe that they were carved by the Olmecs, the mother civilization to the Maya.

The Maya claimed that they descended from the Olmecs. Nobody knows how old the Olmecs really are. They're very confusing to the archeologists, who keep pushing them further back on the timeline, but they can't believe the Olmecs could have done what they were doing in ancient Khemit. If you've ever been to La Venta or Veracruz, there are these great stone heads, carved out of basalt, that look like West Africans.

■ *They definitely have unmistakably Negroid features.*

They are carved well over ten thousand years ago, out of fifteen-ton blocks of basalt. That is the same technology that was used as ancient Khemit. My dear friend, Chris Dunn, has a new book coming out, showing that there were advanced machine technologies in stone carving all around the world, over ten thousand years ago. We see it in India. We see it in Tibet. . . .

■ *I would like to hear more about the information you got from the Mayan Priest.*

What he said to me, again according to his tradition, is that the skulls were not created by the Mayans, but, rather, that they were gifts from their "gods." I always interpret that as one of two meanings: either "gods" meaning extraterrestrials, which many people claim the ancient gods were, or a previously highly evolved civilization, which they venerated as gods. Both could be true!

He was very clear that the Mayans did not carve any large, human-sized crystal skulls. What we know for a fact, coming from tombs, is that Aztecs (a confederation of different tribes; they had conquered the Zapotecs and Mixtecs, who were the artisans of the ancient Mesoamericans—the crystal carvers) commissioned carvers, from other tribes, to carve crystal skulls. Still, these were small, from the size of marbles to the size of maybe a baseball. Francisco was very clear that large-sized skulls were much older, and he believed them to be either from an extraterrestrial source or from a far more ancient civilization.

■ *Can't we also consider that the Mayans and many of these ancient civilizations carved these skulls as effigies of some ancestral memory, like you and I and so many of us seem to hold in our subconscious? Many of us seem to hold the memory of that Circle of Thirteen—a deeply rooted memory of some galactic computer network of crystal skulls, handed down (in my belief) from the Sirians. At least, I believe they were gifts from the Sirians ... but who knows? Who knows how ancient?*

Just because we are discovering these ancient skulls doesn't necessarily mean they are or are not the "Atlantean" skulls. It reminds me of the metaphor surrounding Neo in The Matrix: the perennial question: Are you the One? Of course, Neo is an anagram for "One." It seems people get hung up too much on whether a skull is one of the skulls ... *and then we lose the importance of the experience altogether.*

Yes, I agree. That is ego. So many people who have become aware of the skulls start declaring: "I have one of the original master skulls...."

But as you know, the Mayans were time movers—they were interdimensional walkers. They went into meditational states, using their plant teachers, which we call "psychotropic plants." They could journey to other dimensions and other time zones, and come back, and of course then they would have made effigies of what they had seen and experienced. The Mayans were masters of time and dimensions. They were masters of time—they were time wardens. Certainly, they could journey to those other realities—and to Sirius.

Of course, the Mayans were fascinated with the Pleiades—that is what I was told by Francisco and other Mayan timekeepers; they believed their origins were the Pleiades—that's their home world.

The Khemitians thought home was Sirius and other places—for certain.

So, the Mayans venerated the Seven Sisters. As I mentioned before, you have to understand these things in layers. What we call "ancient" skulls, again, are those that are over a thousand years old. Well, past a thousand years we can go back tens of thousands of years, so again—there are many layers.

■ *Right. If the Earth were to survive another thousand years, and people were to discover the hundreds and hundreds of skulls that we are having carved now, how would they interpret them? Would they deduce that we were all skull worshippers, in this era? It all relates to some ancient memory in our consciousness.*

I think it is important to move past the questions of whether a skull is ancient, or whether it is one of "the ones," and get to the more important question of what these skulls represent to us, in our collective uncon-scious—as well as what they will mean to us, as we move forward.

You're right. It's probably only important to people like me, and Nick, and other archeologists, who are interested in it from the scientific point of view as well, but it is true. As devices and tools that interact with mind and different dimensions and perhaps extraterrestrial con-sciousness, you could use a crystal that was carved last week. You could use any piece of quartz crystal.

■ *I do recommend to people who are buying contemporary skulls, or older ones, for that matter, that they do find out as much as they can about where and by whom they were carved. I know people who carve skulls in Brazil and they are very skilled, and very psychic. Well, you know from working with Marcel Vogel how important the intention of the carver is.*

Yes, Marcel had psychics cutting crystals that were very interesting.

■ *My point is that, of course, the intention of the carver has everything to do with the impact of the skull—from the ancient carvers, to those of today.*

Yes, and the intent of the carvers who worked with Marcel was to create very powerful healing tools, for healing and for raising consciousness. There are spiritual carvers today in Brazil and China who are carving to create aesthetically pleasing crystals, but who do not understand the spiritual impact of that intention. The intent of the carver goes into the piece, as it is being carved.

Francisco shared with me how the Mayans used crystal skulls for healing. He gave us specific techniques that he would use with the Amethyst Skull. The same with Vogel. He taught us that the two principal ways to enter into a skull were through the use of sound and breath. Nick Nocerino corroborated that totally. With his groups, he worked with color, breath, and sound—that was the way to enter into contact with the skulls. Vogel demonstrated this to us: toning, chanting, sound; there were the three key ways to open up the skull and go into it. We started seeing scenes. There's a difference now, between seeing things in your mind, what we call "psychometrizing," and scrying, which means you are actually looking into the crystal and seeing scenes in the crystal itself.

I was privileged to be able to watch Nick work with the Amethyst Skull. The first time I ever saw him work was unbelievable: he showed us great arcs of geological ages in the skulls; great scenes of great cataclysms. Nick said that, in the Mayan Skull, he saw three great ages of cataclysms; in the Amethyst Skull, he saw four. With Max or Sha Na Ra, he could see as many as ten, because they are so ancient—they go so far back, it's unbelievable.

When I worked with the Mayan Skull, I could see scenes of the Maya and the Olmecs. When I started working with the Amethyst Skull, all of a sudden I could see scenes of ancient Egypt. For people to understand that, I need to explain that I have been deeply involved in ancient Egypt all my life, since I was eight years old. Since 1968, it has been my life. I have been studying a lot of ancient Egypt and, of course, I met Hakim, who taught me the indigenous tradition. But when I looked into that skull, I saw scenes I couldn't believe!

I saw the Sphinx surrounded by water; I saw lush green trees. I saw things that proved to me that the pyramids were not tombs for anybody: they served as energy sources. I saw scenes I did not understand, yet I kept them in my memory bank. But I did not understand them until I met Hakim. Hakim told me about how it was in the ancient times, and then I realized:

My god! I had seen these things in the Amethyst Skull in 1983!

I didn't meet Hakim until 1992.

■ *As you know, my book* Where Pharaohs Dwell *makes a direct correlation between the crystal skulls and Egypt, which has never been made before, that I am aware of. I'd like to know your opinion about that. We know there are a lot of links with the Mayans, and Atlantis and that possibility, and Tibet, but . . .*

And Peru!

■ *Actually, I have talked to shamans in Peru, who don't feel a connection to the skulls. One told me it was more Atlantean, while they are more*

Lemurian. . . . Unless you have information that I don't have, I don't believe the Incas have the connection to the skulls, as do the Mayans of Mexico, Guatemala, and Belize.

Is that so? No, I don't have; since I have not been there, I can't speak to it myself. And I don't know any ancient skulls from Peru, so I will agree with you on that.

■ *As I was asking you—I would like to know your feelings about the connection between the skulls and Egypt, since I obviously have a very strong feeling about that and have never heard anything else about it.*

Well, that's fascinating to me, Patricia. Excellent question. Since 1983, when I saw the images of Egypt in the Amethyst Skull, I too became very interested in that connection. What was the ancient Egyptians' use of crystal? When I started looking this up through the Egyptology services, they were totally ignorant of the ancient Egyptians' use of crystal. They said the ancients did not have any quartz crystal and they didn't use any. And I said, "King Tut's tomb shows a fantastic necklace made with various minerals, including quartz beads . . . and they found a quartz crystal goblet in Tut's tomb!"

Egyptologists didn't know that. And we know that they used rose quartz granite in the Great Pyramid: Aswan rose quartz granite is fifty-five percent quartz crystal. That's what makes it so highly resonant and so powerful in sound and vibration—fifty-five percent quartz crystal!

Obviously, they knew about it.

This was one of the most important things I discussed with Hakim when I first met him: "What about quartz?" I asked him.

He was fascinated by crystal skulls. I asked him what the tradition said about crystals and crystal skulls. He said there was very little information, but that he felt that the skulls were universal, and that there was knowledge all around the world that crystals held memory and that, therefore, they held great knowledge. He said he had no problem believing that the Khemitians carved and used crystal skulls. He was open to the idea, even though he had no direct knowledge; neither do I . . . but I am convinced that if there was an Atlantis, it was sharing with Khemit.

If there was a connection between the ancient Maya and Egypt, they would have shared this wisdom. They would have passed these things. So, although we do not have direct knowledge of it, I have no problem believing your channeled information that the ancient Egyptians used crystal skulls.

■ *What's interesting about my encounter with the shaman in Palenque who gifted me with one of his two skulls during that ceremony with my group—is that one of my group asked him where the skull was from. I was so surprised when he said, "Egypt."*

I don't think it is an ancient skull, but he did say it was from Egypt. That was a powerful validation for me, as well you can imagine.

Wow! That's amazing.

■ *It is not clear quartz crystal. Pictures of the skull have been shown to over six geologists, and not one has been able to identify the mineral it is carved out of. One said, "The only thing I can relate this to is a quarried stone used in carvings in ancient Egypt."*

Interestingly, the Mayan shaman calls the skull "Estrella," which means "star" in Spanish, and in Arabic it means "Pleiades."

Nice connection.

■ *. . . So I don't claim that this is an ancient skull, but it is so fascinating that he would hand over this skull to me, declaring that it was from Egypt.*

That is very strong. It is the first information I have heard of a skull connected to Egypt.

We need to remember that crystal skulls record every instant, every sound, every person, every vibration of every person who comes in contact with them.

That is what makes it conceivable that there may very well have been these first thirteen crystal skulls of Atlantis. They could very easily have been the first major intergalactic computer network, and, as you say, that could have remained in the collective unconscious of humanity.

I did get to work with the Mitchell-Hedges Skull in 1988, believed by many to be one of those original thirteen skulls.

■ *Okay, let's talk about that. I would like to hear what you have to say about it, as I too have had a personal experience with that skull.*

It is the one that is most known about, and most written about. It is a beautifully carved, optical-grade quartz—with a movable, detachable jaw. It is comprised of thirteen pounds of pure quartz and fantastically anatomically carved.

■ *So, moving past the question of whether this is a real ancient skull or not, with all the controversy about Anna Mitchell-Hedges' story as to how she obtained it, my question to you is more about what your experience was with the skull.*

Very powerful! It had the highest energy—nothing like that had come into my life, up to that point. I saw Mayan scenes in it. But the thing that hit me the most was its unbelievable beauty. It is so perfectly carved, and that is what bothers me. Why would the ancients have carved something so anatomically perfect? I had conflict with it but then, in the late '90s, Nick announced to me that it was not really ancient. But, because it was so beautifully carved, possibly in Germany (according to Nick, because it is just such a magnificent piece of quartz), it may very well have held the intention of attracting people to it, to be sold.

Actually, what we believe is that the skull was carved in Germany, but it may have been brought back to Mexico and actually placed with ancient skulls, so that the memory held in the ancient skulls could be downloaded into it.

It takes a person with a lot of sophistication and experience with crystal skulls, like Nick Nocerino, or someone who has been trained by Nick and Marcel, as I have, to be able to come to that kind of conclusion. It is not something you can come to lightly. My opinion today is that it is not an ancient skull: that it is old; that it was carved in Oberstein, Germany, in the 1860s. But that does not deny that it is tremendously powerful. It has been with the ancient skulls to be used in the same way they were, and that means it has the energy to be used for healing and communication, the same way the ancient skulls were used.

■ *I recently sat with the skull and found it to be truly amazing. But as I said, I am not focused on whether or not a skull is "real" or the One. I am more interested in my reactions to skulls. I have my own collection of skulls here—others I have sat with in Mexico—even one in Peru, that I found in a back room in a blackmarketeer's warehouse. I always seem to find crystal skulls no matter where I go.*

Again, that is because they call you to them. We believe there is a crystal skull brother/sisterhood (like you talk about, being from Sirius), and certain people, like you, are really connected to the skulls and they call you to them. They do. My final decision on the Mitchell-Hedges Skull is also based on what I saw with Max and Sha Na Ra.

Because of my connection to those two, I feel I can talk about the Anna Mitchell-Hedges Skull.

Those two ancient skulls are crude (they're not too finely carved, not the best quartz), and yet the energy is unbelievable. I know someday soon you have to meet them. I will introduce you personally to JoAnn Parks, who is the guardian of Max.

■ *I've been waiting a long time to meet Max—even from the photo I got that psychic rush through me. Thank you for that invitation.*

There's a whole chapter in the book about how JoAnn Parks, who is the guardian of the skull Max, got involved with a Tibetan Lama named Norbu Chen, very similar to the name you gave your Tibetan skull. Max had actually been obtained in Guatemala from Mayan Priests . . . long story.

■ *Actually, I didn't name it. I got the name from the skull!*

It's unbelievable. My skull gave me the name of "Meketaten," the Pharaoh Akhenaten's daughter.

■ *Yes, they do that!*
    You are the only other person that I know that has made that connection with Egypt. And now I learn that Hakim held a little rose quartz in the pocket of his galabieh.
    We talked earlier about my Tibetan skull, Norbuk . . . I'd like to discuss that again. You saw just the picture of this supposed ancient Tibetan skull and had a strong reaction to it.

Yes, I did. As I said, the first picture you sent was still full of the grit and dust of its patina—it didn't even look like quartz at first. Then, from the second picture of the skull cleaned up, I obviously saw it was made of quartz. It has that ancient, crude style and always the teeth are like an afterthought: the last thing carvers do. That makes it look like it is old and might indeed be ancient. It definitely is not a contemporary skull, carved to be sold.

■ *What's very strange about this skull is that it has an old silver cap on the crown, with seams etched into it, representing the merging parts of the brains. It reminds me of Hakim's teaching regarding the significance of the beetle, or "Khepre," the symbol of awakening.*

It also has silver over the teeth. I have never, ever seen this on any skull before or since.

No, but there have been things done like that. There are skulls out of Mongolia now (actually stone, not crystal), with ancient Mongolian script on the top of them. We have seen things like this, where gold and silver have been worked with them.

■ *Fascinating. I do receive information from the skull that the metal on the teeth and the skull has to do with sound being utilized through the skull.*

I do think Max and Sha Na Ra are the most interesting skulls I have come across—very, very old . . . and again, the idea that they were not carved for aesthetics is interesting. The world-renowned geologist Robert Schoch has looked at Max and he tells us that it is five different pieces of quartz, fused together. That means tremendous heat and pressure were involved. Schoch would say to the anthropologists out there (who insist that all skulls are fakes and modern) that Max could not possibly have been carved by a power tool. If a power tool had gotten into any one of those seams, it would have split the skull back into those five pieces, and it would have cracked and shattered.

It had to have been carved by hand by a master carver: a master shaman. The crystal used to carve Max was chosen because they saw this piece of five crystal matrices fused together, generating tremendous natural energy and power. That is why it was carved as a ritual object.

When we work with crystal skulls, we find that they seem to have a distinctive personality of their own. We have also perceived that they seem to have an interdimensional connection to universal consciousness—not necessarily "extraterrestrial." There are beings, which you know better than anyone, from the fifth and sixth dimensions and above, who are not in physical bodies, but they use devices to connect with us, on our three-dimensional conscious level. Crystals are one of those. They were used as interdimensional communication devices. That's why they get names. . . .

They seem to have a personality and they always seem to be connected to some other kind of consciousness that wants to know what the human experience is—what our consciousness is. So they serve as interdimensional portals.

■ *That leads me to a contemporary question regarding the 2012 scenario. We are almost there. So much points to this date—how do you perceive it?*

You brought this in ten years ago, through your books, which you channeled in 1999—and they're all talking about what is actually happening today. Absolutely. We are coming to the convergence of all calendars and cycles.

I asked Hakim about the date, December 21, 2012, and he just laughed and said, "It doesn't mean anything to me, because I do not recognize the calendar. . . ."

If you talk with genuine Mayan timekeepers, they do not talk of the end of the world. They talk about our moving into a New Cycle: the "no-time."

People say it is so dark now—when is the awakening coming?

Hakim would say: "We are growing every day. What you are going to see right now ... you are going to wake up and know something you didn't know before.... It's happening now. We don't have to wait for December 21, 2012—it is now."

■ *Yes, we are well into the shift now. It is fascinating that the surge of interest in the skulls is happening right in line with the Mayan prophecies, at this very date. And here we are, right on schedule....*

I agree totally. Just in the twentieth century, you have the Hedges Skull appearing, and, since I have been in the field, it is increasing all the time. So, obviously, you are absolutely right. There just has to be a connection between the crystal skulls, the awakening, and the end of the Mayan calendar. The crystal skulls are a part of it; they have always been a part of different civilizations, as far as they were recording and bringing knowledge.

Even Knights Templars had a crystal skull.

■ *Indeed. In the '60s and '70s, we really started rediscovering crystals— it was as if we had to redistribute quartz around the planet, as part of our preparation. To me, the face in the center of the Mayan Skull is a crystal skull and that, in itself, is such powerful information.*

Exactly. Hakim said that to me, when he spoke to me about ending this Age of Darkness and approaching the Dawn. I was amazed that the Sirian Elders were using terms that I first heard from Hakim. He would say the tools will be there for us and that nothing was lost to us; we were just asleep—dormant. Obviously, everything that has served as tools for us in the past is going to be magnified now, as we move to the next stage.

Everything would remain, he said, like the skulls.

■ *And let us not forget that the crystal skull remains the ultimate metaphor of our own immense capacity to use our minds—the crystal brain that we are—to put the pieces together—to remember everything.*

That's very interesting. Yes, the question of: why a skull? Consider—when you cremate a body, two parts do not burn: the skull and the long bones. You can burn a human skull at ten thousand degrees Fahrenheit and it will not melt—it remains. It represents something that remains forever. And where do we store our intelligence? It is stored in our cells, of course, which you are bringing across through your work, but it is mainly stored in the brain. Where is the brain stored? It is in the skull.

■ *Yes, that is a profound observation. The skull always remains.*

We have found three-and-a-half-million-year-old skulls.

■ *Well, the human mind is a computer and the skull is the perfect metaphor—it is all there. There is no question to me that the skulls of the ancients did serve as a computer system.*

I would say **without a doubt.**

■ *So, we agree that the skulls served as a computer system of ancients. Do you feel that we are headed toward rediscovering a group of skulls that will be reunited to perform that function?*

What I got from Francisco was not a specific tradition of thirteen skulls at 2012, but he does believe they will reappear and that the fact that so many people are so interested in skulls is a sign that the awakening is upon us.

■ *The Mayan shamans have been uniting at all the sacred sites, all carrying crystal skulls. The Mayans I am working with are also working with crystal skulls. Something amazing is happening here, right now ... something absolutely magical. At the hour of awakening of all humankind, the crystal skulls are definitely reawakening some ancient memory within us— something we need, to help turn the keys of those locked doors within our collective mind.*

■ ■ ■ ■ ■

Stephen Mehler's book is: *The Crystal Skulls: Astonishing Portals to Man's Past.* Stephen is also director of research of the Great Pyramid of Giza Research Association. You can learn more about his vast knowledge of skulls, as well as his pursuit of the Egyptian and Khemitian traditions, on his Web site: www.gizapyramid.com.

His other books are:
From Light into Darkness
The Land of Osiris

The Cure

■ ■ ■ ■ ■

Jim Humble

A miracle mineral solution in the twenty-first century?

Jim Humble started his career in the aerospace industry, where he quickly became a research engineer. He worked on the first intercontinental missile, the moon vehicle, wrote instruction manuals for the first vacuum tube computers, set up experiments for A-bomb explosions, worked on secret radio control electronics, set up experiments in electrical generation by magnetohydrodynamics, completely wired the first machine to be controlled by computers at Hughes Aircraft Company, and invented the first automatic garage door opener. In the mining field, he wrote four books updating older technology and improving the health hazards for those involved. He first overcame the hazards of mercury and then he invented ways of eliminating mercury from mining altogether. His technology included methods of eliminating chemical leaching, finally using nothing but water for the recovery of gold. Jim's immediate goal is to return to Africa to eliminate all of the malaria in a single African nation in order to prove to the world that it is possible. And he will.... And it is.

■ ■ ■ ■ ■

■ *Dear Jim, I would like to congratulate you for your long journey, and if I may ask you to elaborate a bit on how you evolved to this point, where you are dedicated to healing people of so many diseases through your groundbreaking discovery with the little-known mineral supplement, MMS: Miracle Mineral Supplement.*

I was prospecting in a South American jungle and while I was working there, some of my men got malaria. I tried one of the supplements that I had for purifying water, which I also used for health—and I cured the malaria in such a short time. It was amazing. They were well in four hours.

■ *Yes, I read this in your materials and found it to be almost unbelievable. You apparently add citric acid to a mineral supplement, as I understand it. How does this work?*

Well, actually it is a mineral supplement that has been available in the health stores for the last eighty years, in the United States. It is stabilized oxygen. A lot of health workers know about it and have used it for years and years. I used it in the jungle and my men were well in such a short period of time. The day after I administered it to them, I tried it on two more men who had malaria, and the same thing happened: they were well almost instantly.

I didn't have any choice—once I had discovered it, I had to tell the world about it—that's what I wound up doing.

It took about a year's work on the supplement in order to really make it effective in **all** cases (it wasn't effective in all cases at that time). I sent samples over to Africa, to friends that I had made over there, and we tested it a number of different ways and worked with it, until we found the best way to activate it and until we reached the point where it was **one hundred percent effective** against malaria.

■ *So, you had a mineral supplement to which you added (according to your notes) chlorine dioxide, is that right?*

Basically, the stabilized oxygen is what **creates** the chlorine dioxide. When you add the essential element, citric acid—or lemon juice—to stabilized oxygen, it produces chlorine dioxide. That chlorine dioxide is what you have in your body when you drink it—that's what runs through the body and kills all the pathogens in the body.

■ *Phew! That's a pretty strong declaration! First of all, I imagine you have some resistance coming at you from the mainstream ... but of course I am surely not mainstream! This is Beyond the Matrix—so please let me hear you say that again: it kills all the pathogens in the body?*

Okay, let's say most of the pathogens. I have people take it and increase the number of drops and the amount of lemon juice that they add, increasing it each day, so that by the end of two weeks to a month, a month and a half, they have increased to fifteen drops activated with lemon juice. By then, it has killed most of the pathogens in their bodies. People have hepatitis C—it goes away; diabetes goes; most diseases go away—quickly!

For some it takes a bit longer, depending how deeply the disease has dug into the body. Chlorine dioxide is not the same as chlorine at all—that is totally different. Chlorine dioxide has been used for a hundred years or more to purify water in water purification systems, because it only kills the pathogens. It doesn't go in and make a lot of bad chemical mixtures that could lead to cancer—like chlorine does. Chlorine dioxide is really a different animal than chlorine, and it works totally differently in your body. It is only in the body for a few hours and then it disintegrates into table salt, so it leaves nothing behind to build up and later cause side effects.

■ *Your information sheet says it also has cured cancer. As that is such a huge concern for our growingly toxic societies, I would like to ask you to explain more about this possibility—a cure for cancer?*

A woman just called me from Australia about a week ago. She had lung cancer and the doctor told her she had about two weeks to live. She was bedridden, although she had been up in a wheelchair, for about ten minutes a day, just to escape the bed.

Her family decided to give her a birthday party, right away, because they didn't feel she would live to see her birthday. Someone suggested she start on MMS—her doctor, actually, brought it over for her—and eight days later she was up walking around. She hadn't been able to walk for a couple of years. At the end of twelve days, she drove down to the lake and walked along the lake. That day she held a seminar. She is now back, teaching school.

■ *That is utterly amazing. Do you have any clinical trials to back up your claims?*

I did clinical trials in Malawi for malaria. We did clinical trials in a prison there. We had a separate company testing the blood—they wouldn't let us even get near the blood, before or after the treatment. There was a one hundred percent recovery of all patients treated. The government thought we must have been mistaken, so they set up their own tests and they got the same results—one hundred percent recovery of all patients treated.

■ *One hundred percent?*

From malaria, yes ... one hundred percent! When we treated AIDS, we did so with intravenous injections. There we got a ninety-nine percent recovery.

■ *Do you have documentation to back this up, Jim? I'm just playing the devil's advocate here, because I know people are going to ask this question straight away.*

We do have documentation for malaria treatments—a person can visit or call the government over there, if they want to get that—but not for the AIDS. We never got documentation, because we didn't have the funds to pay for blood tests. We didn't test them. The people just felt good. They got out of bed, got dressed, and said: "I'm going to go home to my family."

And that's what they did.

So what we can say is: they didn't die of AIDS. They all went home, feeling good.

■ *That is earth-shattering information. And what are you doing now, Jim? As I recall, you are back in Africa, intent upon eradicating malaria there altogether. Is this right?*

Right. The sales of my book will hopefully pay for eradicating malaria in one country. Hopefully, that will be Malawi, because I know people in the government and I have friends there. There are twelve million people in the country and about four million of them have malaria. I expect to be able to treat all of them in about eighteen months, and it will be paid for mostly by revenue from the book.

■ *How are you going to treat four million people in such a short time?*

I'm getting some donations and I will have enough income with the book sales to be able to hire people—really good people—for forty dollars a month. I'll get a couple of cars, hire some people, train them how to use MMS—and then send them out to the villages, training the chiefs in the villages. I've already done this. The chiefs were totally willing. They and their medicine men learn how to use it; they then treat the people in the village. So we believe we can reach everyone in that time frame. Of course, we'll be manufacturing MMS right there in the country. I can manufacture it for less than a penny a dose. It only takes two doses of MMS to cure malaria.

■ *What? Two doses?*

That's right—two doses.

■ *We need you to explain this to us. You train people on how much lemon juice to add to the supplement . . . you squeeze a lemon, add it to the supplement, and bingo! A person is healed of the dreaded disease of malaria, after two doses of this cure, at a cost of two cents? It's that simple?*

It takes fifteen drops—I use eighteen usually—of MMS and one hundred drops of juice (I actually use citric acid, rather than lemon juice, as I make one hundred gallons at a time). Citric acid is just like lemon juice; lemon juice is ten percent citric acid and that's what I use. I put one hundred drops in there, wait three minutes for the chemical reaction, put it into a little pineapple juice, and I give it to the person. That's a pretty heavy dose.

I wait an hour and then give a second dose. By the end of four hours, more than ninety percent of the symptoms are already gone: the fever is gone, the headache is gone, the shaking is gone, the nausea, the aching in legs and arms: all gone—in four hours. If the patient is bedridden, he gets up, puts his clothes on, and goes home.

■ *Amazing. It is just too exciting for words.*

I've personally treated over two thousand people and then the people I have trained over there have treated over seventy-five thousand. We have plenty of evidence that it works. I don't know of anyone who was not healed. Of course, there are people who are sick with other illnesses as well, so they are not always instantly well, because the other diseases often take a few extra days—even a week. In any case it's very, very fast and very exciting. No matter how often I do it, it is so exciting to see someone get well so fast—it is just great.

■ *It's beyond great! What an incredible humanitarian moment in life! To know that you are doing this work for so many people, and to see these immediate recoveries, is just hard for me to imagine.*

What about this obscure new disease, Morgellons? How does this cure affect this disease?

People with Morgellons have all benefited from MMS. It is, unfortunately, not as fast as with malaria, and some of the people call me up and say, "I have my life back and I'm doing okay."

Nobody has called to say that the disease is completely gone. The ones that say that they're doing well and functioning still feel some parts of the disease working in the body. So it hasn't totally cured that, but we still have a lot of research to do.

I have been working and testing it intravenously, activated. No one has ever done that before. I've been trying it on myself, to see what it does for the body.

■ *You're injecting yourself? Jim, you are an amazing man. What are your reactions to the injections?*

Well, I got freezing cold about ninety minutes after injection; I had to cover up with a bunch of blankets for a while. The next day, I was able to take more intravenously than the day before. It's cleaning something out of my body! It's cleaning my lungs out. I had a noise in my lungs when I breathed—that noise has been there for ten years—all of a sudden, just a couple of days after that first injection, it was gone. So, I was clearly on to something deeper there.

There are people who have been using injections with the same chemical I'm using—there have been more than a hundred thousand injections in hospitals and clinics around the world, using the sodium chloride in about the same amounts that I use.

The difference is that I activate it, but they don't—they just use it straight. The activation increases the amount of chlorine dioxide by huge amounts—maybe a hundred times more.

■ *So it is the chlorine dioxide that is the inevitable healer. That is what I hear you saying, right?*

Chlorine dioxide is what does the work. It is a killer. There's no nutritional value in it—all it does is kill things. It is an oxidizer—the most active one known to man ... but it is still a weak oxidizer, in the sense that it can't attack healthy cells, or friendly bacteria. It won't attack them, because they are too strongly put together to be oxidized by chlorine dioxide. Otherwise, it is a great oxidizer and killer of pathogens.

■ *Sorry, Jim, I'm a bit confused. How is it that cancer, which is so overpowering in the human body, can be killed by this chlorine dioxide, yet a healthy cell cannot?*

When a cell becomes diseased, it becomes weakened. Cancer cells are much weaker and they have a different construction in the cell walls than healthy cells. I'm not sure of how it kills the cell. It may go through the cell wall and get into the nucleus of the cell—killing the nucleus. But it does kill the cancerous cells. They break loose and wash out of the body. And in the process of washing out of the body, they release a certain amount of toxins in the body—so a person can get somewhat sick and nauseous while treating it—they can get diarrhea and that tells you that it is working.

It will kill the cancerous cells. How it does it, we're not sure yet, but the fact that it does it has been shown to happen in hundreds and hundreds of people. These people are calling me and telling me it is working.

■ *So we don't really know why this is working.*

As far as cancer, no—we don't know how it kills the cancerous cells. With the other pathogens in the body, it is pretty well understood, scientifically, how chlorine dioxide kills them. It has been used in hospitals to purify the floors and other utilities. It's used in slaughtering houses to purify the meat. It is used throughout the world in many water purification systems. It is just as well known how it kills pathogens in water. It's selective. It only kills the pathogens in the water purification systems and it doesn't attack the equipment either. It is a weak oxidizer. All pathogens are anaerobic-type bacteria and viruses—they all are attacked by it—but the healthy organisms that you want in the water aren't attacked.

■ *Let's say that by the time a person gets the information of this mineral, he or she will have already gone the route with chemotherapy and radiation. Would one still be able to be healed by this treatment?*

No, the more they have been treated with chemotherapy and radiation and surgery, the less likely that the MMS is going to help them. It sometimes helps, depending on how weak they are and what they can stand. It does its thing—it starts doing what it does, but you have to be killing those cancer cells off faster than they're being generated.

When you kill a cancer cell off, it releases a certain amount of poison into the system. What we try to do is to keep people taking just enough drops that they aren't getting sick—that is, one drop away from experiencing the nausea and other effects of clearing the cancer out.

If the cancer cells are growing faster than you can kill them off, then the person is probably not going to make it through the MMS cure. Most of the time (like medical people say), the sooner you catch cancer, the better chance you have of destroying it with MMS.

■ *Do you recommend that people self-treat with these supplements?*

Yes, if they can. Yes! There are thousands of people out there with cancer and not that many doctors who are willing to use it at this point. There are right now thousands of people self-treating ... more than ten thousand bottles are sold each month in the United States. All kinds of people are using it, for all kinds of different things. Most are getting really good results. If a person has been radiated, and treated by modern medicine too much, it can be ineffective.

■ I am so thirsty to know more—especially since you are talking about most pathologies and not just malaria. What about prevention? Do you recommend MMS as a preventative path?

Yes, absolutely. I recommend you start out by working your way up to quite a few drops—to the maximum limit of fifteen drops. Then, if you take fifteen drops three times a day, for a few days, you will know for sure your body's pretty much cleaned out all the pathogens—unless something is really, really extra deep. You clear your body of all the pathogens and you will know it! You feel different. After that, you should take maybe two doses of six drops each week, of course activated with citric acid. That should kill any new pathogens that get into the body.

I have people who get the flu—usually a couple of doses of this will kill off the flu in about twenty-four hours. A cold usually goes in an hour or so. Sometimes it takes a little longer—maybe two to four hours, but normally you can get rid of a cold in that amount of time. Sore throats as well—and of course if you use it on your teeth, it changes the health of your mouth. I don't care how you have been brushing your teeth and taking care of your mouth. If you use this in your mouth for a week, you'll notice the improvement in the overall health of the mouth: your gums will get stronger, your teeth will be tighter in place, and you won't ever have any more bad breath. Your mouth will just feel stronger.

■ Jim, do you sell this mineral on your site?

No, I don't. I don't make any money from the sales of it. Some of them donate money to me, but I decided that, if I was going to talk about it in my book, I couldn't sell it. So I do not sell it.

■ Is that because you thought you would get shut down?

Yes, the FDA would be very hard on me if I were selling the supplements and the book. Also, I have a better response from people around the world, since I am not selling it. People know I'm not in this for the money.

■ *Obviously that is very important to people—eliminating the suspicion that you have any financial motivation behind the work you have taken on.*

So, the FDA is "allowing" you, for the moment, to go about your business, telling people about this apparent miracle cure. But Big Pharma is now going to take over all mineral supplements, as you surely know. What will this do to your work and your freedom to expound on the wonders of this incredible treatment?

They will stop me. I'm sure they will stop it. But we have done things that can slow them down. I call it a "miracle mineral solution" "water purification drops" on the bottle. There are no claims other than that. The claims are all in the book. The bottle only talks about purifying water. Nobody can object to water purification drops.

■ *Well . . . that's to be seen.*

Well, yes, when the time comes, they will come and they will stop it. There's no doubt about it, because it will be reducing the income of the drug companies and they're not going to put up with that one little bit.

■ *Especially now that we have all these new emerging pathogens, viruses . . . and all these expensive drugs, like Tamiflu, are being presented as the sole cures for them.*

What about the bird flu? Do you think it will be effective on this as well?

I think so. We haven't tried it on bird flu yet, but it definitely kills anthrax! Anthrax is one of the worst pathogens around.

■ *It kills anthrax?*

Of course! That's what they used back when they had those anthrax incidents. They used chlorine dioxide to kill the anthrax. They used the same stuff, sodium chloride, to generate it with. The only difference that I do is that I put it into the body. But I use the same stuff. Actually in the same concentration—two to four parts per million of chlorine dioxide kills anthrax and kills almost any other known pathogen known to man. I assume it's going to kill bird flu and all the rest of them.

... And those people who are creating the diseases, like Morgellons and Lyme disease, and a whole bunch of other diseases that trace back to the laboratories, I'm sure they're not going to be very happy with MMS, because sodium chloride kills all of them.

■ *Have you been intimidated yet?*

No, other than a few doctors saying that I am doing wrong ... most of the doctors are on my side, even those you would think might not be. But there have been doctors who told me that I'm doing wrong and that I should wait until they spend a hundred million dollars to prove it out. That's the only way we can get approval—through the standard process, and that requires a hundred million dollars.

■ *So, what can you tell us about this book? Do you provide ample descriptions of how to work with this substance? What kind of practical information do you give people in this material?*

I give complete disclosure in the book. Not only do I tell you how to take it and how to use it for various diseases; I tell you how to make it. Not only do I tell you how to make it in your own kitchen, but I even have a chapter on how to manufacture it in your own kitchen. So you could make two to three hundred bottles a day, just following my instructions. I have made every attempt to tell people how to make it and how to use it, so that it will never be lost. The book is out, around the world. No matter how they try to stop it (and they will in the United States), the information will still be out there, in the books, and it will still be around the world.

■ *I think it is important to reiterate that the money generated from this book is going toward healing four million people in Africa—a very noble quest. Tell us—what is your ultimate goal for Africa?*

Well, once the world sees that we just about cleared up an entire country, there will hopefully be people who will help and donate the money to help clear up other countries as well. Other countries have shown an interest, but we're going to need a lot of money to clear up Africa of most of the diseases that are there. Once we do that, we won't need to put up as much money, as we need to now, in Africa.

The free countries of the world are putting billions of dollars into Africa. Nobody really knows how much! Every city in Africa has a nonprofit organization helping out, but it's not nearly enough, because there is just so much disease there.

Once we eliminate most of it (and that's if they allow us to), a whole lot less money will have to go into Africa. Five hundred million people get sick from malaria every year—that creates the majority of poverty there. Once these people are able to work again in the harvesting and planting seasons, the amount of poverty will be reduced ninety percent. This is going to change things there—it will change the amount of money needed to keep the people in Africa going. It will change the world.

▪ *I wonder really about the "powers that be" and whether their true objective is to really heal Africa and its people, or whether it is to perpetuate their own agendas.*

I think that has been the case ... they've just perpetuated it. Even when we spoke to Bill Gates, he said that, as soon as we got FDA approval, he would talk to us about a grant, but he knew that we would never get FDA approval, so he didn't have much to worry about. Of course, I have applied to all the big organizations and they either ignore me or they send me a letter of rejection.

At least Bill Gates has sent me letters of rejection, as opposed to **nothing.**

▪ *Tell us more, Jim, about the actual treatment process itself. You're in Africa, people are lining up, some of them know they've got malaria, some of them affected with who knows what outrageous disease. How do your people know what they're treating and what doses to be given? What is their overall sense of what approach to use?*

The approach for malaria, at least, is that you have to give them real heavy doses, to kill the parasite off right away. When you see the person, you recognize malaria. It doesn't take long to recognize it— anyone in Africa can. You get to know it really quickly. You give them a really heavy dose, and then some of the other people, who are not as sick as the ones with malaria, might get a lighter dose.

When we go out to a village, we give everyone with malaria a pretty heavy dose, and the others, who are just complaining of not feeling well, we do give a slightly lighter dose. The people with malaria were getting eighteen drops and the others were getting about twelve.

Some of them would experience the nausea and vomit right away, but we would come back the next day and most of these same people were well. Most of them came by to tell us they were feeling good—healed.

They would get in a long line in the village and tell us what the problem was; the person giving out the drops would just mix the drops right there and provide the doses. We had to analyze things right there on the spot, because we didn't have a lot of time for each person. It worked beautifully. If some got sick, it was because they'd get the vomiting or the diarrhea, but they got better... which is what mattered.

■ *Some people would say this is a kind of "guinea pig" approach. What would your response be to that?*

The people there didn't think so. They were happy to be well the next day. There are a lot of people over there lining people up, giving them pills and various shots, and whatever. If you want to talk about "guinea pig" situations, the World Health Organization gives four to five million shots over there of what is supposed to cure or prevent various diseases, and most of the people wind up sick or sicker. It has been pretty well established that the WHO did quite a bit of damage over there. Of course, nobody's willing to admit it, other than the people who really understood what they did. It's been written about in books and all kinds of reports....

■ *I guess the ultimate testimony is from the people who say to you, "Thank you—I'm healed."*

That's right. The ultimate is the people coming back the next day, gathered around us, bringing their family members in, everywhere we went—people were excited about it. The ones who went through the vomiting ... well, we told them they might and they were happy to do that cleansing, since they knew that it meant the cure was working. There were hundreds of people, in the villages, and when we were with the missionaries, the people would come to the churches, where we'd visit the churches and a lot of people would be there with malaria.

A lot of people walk around with malaria. They're too sick to do much work, but they can at least move here and there. All of them arrived at the church to be treated. Most of them would be feeling pretty well after treatment and already feeling much better before they left to go home.

■ *What about animals, Jim? Can animals be treated as well?*

Sure! Lots of people are treating their animals with it. I have people calling about their animals and I've mentioned in the book that I used to use three drops for each twenty-five pounds of body weight, but you start out at a smaller amount, so they don't get the nausea and diarrhea. If a person gets sick from the treatment, he's not going to want to continue with it, so in the U.S. we start them off with one drop, usually, and slowly increase, so they stay below that sickness level. If they are getting slightly sick, though, they know it is working and they are killing something off in their body. Animals ... you can pretty well tell how the animal is coming along as well.

■ *What about the average person, who may not be dealing with a terrible illness, but has a general sense of malaise, disharmony, chronic fatigue, etc.? Would you recommend the supplement and, again, how would they dose themselves?*

Yes, people with candida and chronic fatigue should start with one drop the first day, two the second, three the third, etc., and, sooner or later, they're going to go to enough drops (always fifteen maximum) that they begin to feel the nausea. So, for example, if you take eight drops and you experience the nausea, you reduce back to six drops, for two or three days, and then start back up again, until you reach fifteen drops. Most adults go to fifteen drops; for children, we're talking about three drops for maximum dose. That's three drops for every twenty-five pounds of body weight, but you don't start with that—you start out at one drop.

■ *And when you reach fifteen drops, do you take them just one day only?*

When you get up to fifteen drops and you've gotten to fifteen drops without feeling much more than a little nausea on the way up to fifteen, I recommend that you take fifteen in the morning, another at noon, and another at night. For the most part, all of your pathogens are gone by that point and so adding another fifteen you won't notice anything. You should be able to take fifteen three times a day for a week or less than a week (around four or five days), without noticing any discomfort or any particular feeling from it at all. That's when you can be pretty well sure that you have killed most of the pathogens in your body.

■ *Even if you don't know what pathogens you've got—they're dead!*

Right. People come here and we train them. We have them take six drops right on the spot, wait an hour, and then we have them mix their own—so they get two six-drop doses, one hour apart. I'll give you an example: a lady came and her hand was totally paralyzed and so was her right foot, from the ankle down, and from the wrist down on her hand. She had not been able to move them for two years. Her husband had to hold her on to a walker, so she could walk in the door.

We gave her a six-drop dose and waited an hour (she had a bad pain in her sciatic nerve and a pain down her leg when she came in) and, after the first half hour, she started noticing the pain going away. At the end, she started getting a little feeling in her hand ... within one hour. She took the second dose. By the end of the second hour, she was moving her hand normally and she could move her fingers and her wrist. She took her shoe off and she was moving her foot as well. All the pain was gone, from down her leg and back.

About the best I can tell you is that the pathogens and bacteria and viruses form along your legs and in your knees and different parts of your body, and they eat away at them until you start feeling the pain. This woman had those pains for twenty years! Two hours and they're gone!

■ *Two hours against twenty years! I'd say that's worth a try....*

Jim, your work is truly remarkable. How would you sum up your message to the world?

I would just say, "**Try it!**" What can you lose?

Thousands and thousands of people have so far. I don't guarantee to anyone that it is going to heal what ails you ... but it is worth a try. I don't care what's wrong with you: give it a try. I first thought it was only going to work for malaria and AIDS, but now I see it works for everything. It's uncanny. I continue to be amazed every day.

■ *Jim, you are a Renaissance man, out there, healing the world. You have all of my admiration and you can count on me to help get your word out to as many people as I can ... to help you get the support you need.*

■ ■ ■ ■ ■

You can read more about Jim Humble and his work at:
www.miraclemineral.org

The book is:
The Miracle Mineral Solution of the 21st Century

The End of Money

■ ■ ■ ■ ■

Thomas H. Greco Jr.

Thomas H. Greco Jr. is a community and monetary economist, educator, writer, and consultant. He is a former tenured college teacher who has spent more than thirty years studying and writing about ways to achieve greater harmony, equity, and sustainability through business and economics. And, most of all, he is a visionary of a world where humanity can create peace and build, rather than destroy....

■ ■ ■ ■ ■

■ *As we observe the breaking down of a lot of those structures that give us a sense of security, what does the future hold? This is an exciting and absolutely urgent subject that the entire globe is confronting at this hour of our evolutionary transmutation.*

Thomas, I invite you to go to the absolute nut of what you are proposing in your book! Please tell us a little bit about your background, and how you came to write your latest book, **The End of Money and the Future of Civilization.**

I started investigating the money and banking problem back thirty years ago. The reason I focus so much on this is that I realize that it is one of the central mechanisms through which power and wealth are concentrated in the hands of a few people. I started writing about it twenty years ago—my first work was entitled *Money and Debt: A Solution to the Global Crisis.* In that book, I outlined the nature of the dysfunctional money system and put forth a few principles that I thought should be applied to developing a more functional one. A few years later, I wrote another book, *New Money for Healthy Communities,* which looked at alternative exchange mechanisms that have been used historically and which were being used at that particular time, 1994.

■ *I would imagine you were able to pull from the global situation current now to be able to enhance what you have already written. Let's talk about this global collapse, which I feel is manipulated—but of course I am not an expert as you. But I am intrigued that you go into the process of the New World Order and how this is all being played out, as you say, to the benefit of a handful of the financial and power elite. Can we get an idea of your knowledge of how the whole thing has been and is being manipulated by and for this very elite minority?*

Sure. We've been kept ignorant of the mechanisms by which money and banking work. Very few people really understand it. There is a lot of obfuscatory rhetoric that is intended to keep people ignorant. When you look at the recent hearings in Congress, where certain congressmen, like Alan Grayson, have been questioning the leaders of the Federal Reserve on what they have been doing, and they have been reluctant to reveal that, which is an indication of how far out of control the system has gone.

■ *Not only reluctant—they are simply refusing to answer . . . and they seem to know that they can!*

Yes, they are indeed refusing. It is a very arrogant stance that they're taking—it's like they've carved out this realm for themselves and they are beyond the pale. It was sold to the people on the basis that we needed an independent central bank so that money would not be manipulated by political interests. But in actuality, it has enabled the manipulation of money by political interests, so what we have at this stage is a central banking debt/money system that has spread all over the world. Virtually every country in the world has a central bank and the collusion between bankers and politicians amount to this: the bankers get to lend money at a rate of interest (money that they create through a few bookkeeping entries) and the politicians get to spend as much money as they want, without regard to tax revenues, and that money will be monetized by the central bank, if necessary.

In essence, we have these two parasitic elements in the monetary system: interest and inflation. Inflation is the debasement of the currency through the monetization of government debt. Government debt creates new money when it is monetized, but there are no additional goods and services going into the economy, so that money represents "empty dollars," so to speak. It forces vendors to raise their prices in order to protect themselves from this legalized counterfeit.

■ *So we're talking about legalized usury?*

Yes, usury is built into the system. What goes by the name of "interest" is often really just usury, charged by the creators of money for the use of that money. Basically, we have allowed the bankers to privatize the "credit commons," as I call it. We give our credit to the banks and we beg them to lend some of it back to us ... and they charge us for the privilege.

■ *It's outrageous. How do you see this unfolding as we watch the government taking over the banks and taking over the private sector? Even though we hear the occasional cheerleading messages from the administration, saying that it is getting better, it really isn't—is it?*

No, I don't think it is. It has a long way to go before it gets better. I think the improvement is going to be highly limited, as long as we have these current banking structures. If we are going to see any significant improvement, I think we are going to have to create alternative exchange structures and alternative mechanisms of finance.

You mentioned the government taking over the banks. That is really not the case: it may look that way on the surface. But really, if you look at who is in charge ... both the current Secretary of the Treasury in the United States and the previous one were executives of Goldman Sachs, probably the most powerful financial institution in the United States and one of the most powerful in the world. The question is: whose interests are they serving? What is the Treasury doing that benefits the people? What is the Federal Reserve doing that benefits the people?

It is really primarily for the benefit of banking interests. With the bailout of the banks, we have the shift of private debt to the public sector. The purpose of the central bank is to enable the monetization of government debt, which means inflation, to enable the banking cartel to charge interest on money that they create, basically, out of thin air, by making some bookkeeping entries! We have this arrangement that has allowed a small group to capture control not only of the economy but also political control as well. So, we have the trappings of democracy, but in reality, we have a world that is run by a small group of elite, who control money, banking, politics, finance....

■ *When you say "the trappings of a democracy," what do you mean? It sure isn't looking like democracy to me.*

Well, they give us the option of voting every few years for people who are pre-selected.

■ *Yet, we are realizing how that is manipulated too.*

Yes, of course it is.

■ *So what we have is the illusion of democracy.*

Yes, "illusion" would be a better choice of words.

■ *This is such important information for people, because one of the ways that people are stirred into obedience and allowing these events to occur is on the grandeur of terms such as "democracy" and "patriotism"—nation-alistic thinking—but in reality, we are witnessing the New World Order taking hold.*

It's true. The political powers have been talking about it for decades now—the question is what do they have in mind when they talk about a New World Order? They have been very thin on specifics as to what it means.

■ *Of course, they are deliberately vague about their plan, so that it is in place before people start asking the right questions.*

Right. They like us to think that there is no choice. That's what they told us about globalization: that it has to be done, that there is no choice, and that it has to be done the way they tell us to do it. Basically, the Old World Order, as I can discern it, is based on: a belief in scarcity; an immutable and sinful human nature; violent conflict; command and control methods of imposing order; and on elite rule. The New World Order is just that, based on the same premise.

■ *So there is nothing new to it, is there?*

Nothing at all. It is further in the same direction that we have been going for over three centuries. What we have to do is create a convivial new world order based on beliefs in abundance and sufficiency, on collective intelligence and community self-government, potential for human improvement, nonviolent conflict resolution, and distributive justice. These are the systems I am trying to promote.

The characteristics of this kind of convivial new world order would be community-based, democratic structures: more cooperation; equitable distribution; a greater sustainability; a regenerative and sustainable economy. It would be supportive of universal human improvement—not just making things better for a few at the expense of the many. And it would be a peaceful world order, where we cooperate for the greater good.

■ *Do you see this as a possible reality? How do you see this taking hold?*

Yes, I absolutely think it is a possible reality and many people around the world are working toward it right now—maybe not describing it entirely in those exact terms. We have the Green movement, the decentralist movement, the bioregional movement, the human potential movement. We've got the Transition Towns movement. We've got many alternative exchange mechanisms and community currencies that are popping up in places all over the world now, and with the current global financial crisis, people are really starting to sit up and take notice.

The flaws in the dominant system are clearly being revealed and people are applying themselves and their energies to creating these alternate structures. I liken it to the metamorphic process of the caterpillar into the butterfly. This may be a bit speculative, but I think it is a good vision for us to take hold of and work toward.

■ *We are very close to 2012. Whether or not you believe in or embrace the Mayan calendar, certainly it has been proven to purport that this is the end of a galactic cycle. Some people believe we are facing Armageddon; others believe we are at the threshold of a New Dawn. Regardless of one's slant on this, it is unquestionably a time of galactic change, which is very exciting. The Mayans did predict that the world would, in a sense, crumble, so that it could be reborn into a higher vibration. In fact, we are seeing the crumbling of a lot of institutions that we thought were infallible.*

I don't know much about the Mayan information. I have looked a little bit at it, but I don't have a very deep understanding of what was being predicted. But factually, we can see that our institutions are crumbling across the board. Our health care system fails to deliver health care; our education system fails to deliver adequate education; our economy is based on the continual imperative of growth, which cannot continue indefinitely. At some point, some things have to mature.

Ecologists study insect and animal populations, and they can see very clearly that when a population grows exponentially, it has to eventually level off to a steady state, or else it collapses catastrophically. I think that is the kind of change that we are going through in our world.

We have a multidimensional mega-crisis, and that crisis is largely being driven by the compound interest formula, which is built in to our monetary and banking system. Compound interest is an exponential growth function; that means that it grows at an accelerating rate. So, if we are going to manage the transition to a steady-state economy, where we are able to subsist on the resources that we have, we are going to have to change the way we do money and banking. That is what my books are all about.

■ *The title of your book is so compelling. It really does invoke a sense of a whole new way of being on this planet. Can you give us your insights into what will replace the current economic and banking structures? Moreover, can you give us further insights into what you are proposing in your book?*

People often ask me, "What do you mean by the end of money? Are we really going to have an economy without money?" The answer is: yes, and no! Money has been going through an evolutionary process—we started with commodity monies—gold and silver. Prior to that, we had all kinds of commodities that served as exchange media: hides, nails, bullets, sugar, salt—all kinds of things. Those commodity monies were generally useful things that people would accept in trade, knowing that they could pass them on later for something else they might want. For example, when tobacco was used as money (as it was in U.S. colonies prior to the Revolutionary War), people who had no use for tobacco would still accept it, because there was a general demand for it. They knew they could exchange it later for something they did want.

We evolved from commodity money to symbolic money, which was paper money that was simply a warehouse receipt for commodities on deposit somewhere—typically gold and silver coins on deposit in the banks. Later on, banks discovered that they could create money based on people's credit—so they started issuing bank notes, not only based on gold and silver on deposit, but based also on other assets, like a mortgage note or a signature note, which one might sign and ask the bank to take as collateral for a loan.

This creation of "credit money" liberated us from the limitations of commodity money, but it also opened the door for greater abuses, because it was less transparent. Bankers then started to issue paper money on the basis of collateral that had questionable value or no value at all, and on the basis of long-term government debt that might never be repaid. That basis of issue of money basically debased the value of all the money already in circulation—that is the process called "inflation." What we need to do is reclaim the credit commons and to take control of our own credit in our own communities and help allocate it according to our own ideals and the values that we ourselves choose.

What I am talking about is a process called "mutual credit clearing." There is a history of this—the credit clearing process is nothing new. What I am talking about is the general application of it in credit communities around the globe.

■ *How do these communities free themselves from the dictates of the Fed, who is in your pocket for a huge percentage of your personal income and personal wealth?*

The first thing we have to do is to liberate ourselves from dependence on the dollar, on the euro ... the yen—any of these political currencies that are all managed pretty much in the same way by their respective central banks. The way we do that is by organizing these credit circles that I mention.

■ *So, you are talking about communities that are creating their own currencies and that are essentially declaring themselves separatist from the federal system?*

No, there doesn't have to be any declaration of separatism. It is simply an offering of an alternative way of paying for purchases.

■ *In other words, a barter-based system?*

That is the way people generally think of it, but it really isn't barter. Barter is one on one—I have something you want, you have something I want . . . let's make a trade. That's barter.

■ *Right, but some of these communities based on a barter trade system will give you credit: example—I have ten bushels of wheat; I get credit for that, to be used for the purchase of some other commodity or services.*

The credit originates only when you sell that to somebody else in the system. The best example that we have of this is a credit clearing system that was organized in Switzerland, during the Great Depression. It was called "WIR," started by the small and medium-sized businesses that existed in Switzerland. At the time, Swiss francs, like other political currencies, were scarce. Banks were restricting the supply of credit to the commercial and productive sectors, so these small business people got together and tried to figure out "how can we continue to do business with one another, even though we don't have any money?"

They all had products and services to sell, but there wasn't enough official money in circulation to mediate the exchanges. So, they decided to just set up a system of accounts: when you sell something, your account will be credited; when you buy, your account will be debited. When you sell, your account increases; when you buy, your account decreases.

They organized this exchange system, called WIR, and it has been operating for more than seventy-five years. It has been thriving over that period of time.

■ *How do they escape the clutches of the government—saying they have a viable income from products being sold—which wants its tax retributions?*

Taxes are a separate issue. They still have to pay taxes ... but that is not the point. The point is they continue to do business without having to use Swiss francs. So they liberate themselves from the political money system to that extent: that they simply use their own goods and services to pay for the goods and services that they buy. That's what mutual credit clearing is all about.

Recent statistics available from WIR indicate that they have about sixty thousand members nationwide conducting business of around 1.3 billion U.S. dollars, every year. This is a significant national entity.

These kinds of entities exist in other parts of the world. They are called "commercial barter exchanges," even though "barter" is a misnomer. They are more correctly called "commercial trade exchanges" and they enable business-to-business transactions, using this credit clearing process. There is no reason why we can't set up these credit clearing exchanges on a more generalized basis, to involve employees and consumers and all kinds of businesses: from the retail, to the wholesale levels, to the manufacturers, down to the basic commodity producers.

■ *So, basically, this type of system does not allow the possibility of getting into debit status, because you don't accrue debit if you don't have credit in your account, is that right?*

Not so. You are allowed a line of credit, which means you can buy before you sell. Some issuers that have established themselves as creditworthy will be given the privilege of having an account that goes into the negative.

■ *Who establishes that person's credibility?*

The community itself. That brings up another question, as to how these exchange systems should be organized and managed. They can be organized as cooperatives or as private, for-profit businesses. The commercial barter industry is largely organized as commercial, for-profit businesses. But they still provide this credit clearing service, which is a valuable service for their members.

■ *What prevents this arrangement or organization from falling prey to corrupt management from private interests that want to avert or divert to a more conventional system? In other words, certain individuals achieve a decision-making position, which empowers them to determine whether someone can or cannot have the privileges you mention. Soon, these individuals have too much of a voice and power as to the direction the community is going.*

Yes, this is a common problem and we have seen it, over and over again, in human history—where the hired hands usurp the management and take over the privileges and rights of the owners. The reason why we currently have the situation we have is because people have not taken responsibility in asserting their own power to assure that things are done properly. What we have now, as part of this transition process, is people taking responsibility and learning what they need to know—asserting their power. I talk about money power being in the hands of the people because, as I said before, it's our credit that we give the banks and then beg them to loan some of it back to us.

We can allocate our credit directly. But how do we organize these systems that will prevent the usurpation of power and a small group taking over? That's a whole other question entirely—perhaps the best way to do this is to organize these systems as cooperatives or, if they are commercial systems for profit, then we have to be sure that they maintain a level playing field in competition in the industry.

■ *It does seem that it is just waiting to be infiltrated by interests that would not want this sort of power, in the hands of the people, to proliferate.*

Well, yes, we've seen that in the past. There have always been vested interests that didn't want the slaves to be freed, and others that didn't want women to vote—those are things that we have to contend with.

■ *... Hoping that we are intelligent enough to recognize that if we don't make changes, everything really will collapse.*

Yes, and it is important to stress the fact that by making these changes, everybody is going to be better off. The system, as it stands, is simply unsustainable, so those who think they are in charge are going to be in charge of nothing at all, ultimately. So, we need to make the changes that empower communities and help us to transcend the dominant monetary paradigm.

■ *It seems to me that humankind almost needs to be pushed to the brink before we rise to the occasion and rally, understanding that we are at the verge of some kind of disaster and that this urgency is finally awakening us to the reality that the time of enormous change is upon us.*

Yes, crisis always brings opportunity and as you say, people don't react until they start to feel the pain and see that there is a necessity for change. We have within us differing impulses: we have the conservative impulse that wants to keep the status quo and we have the change impulse, which wants to explore something better and more interesting.

■ *What is surprising is the "conservative impulse," as you call it, wanting to keep things as they are, considering that, as you just said, they will soon be in charge of nothing, when it all comes down.*

True conservatives are not the ones who are trying to keep the status quo. Take Ron Paul, for example. He has been trying to sound the alarm on this money system for ages. He's starting to get an audience, and people are starting to pay attention to him.

■ *He's far from a conservative, though!*

He is a conservative. A true conservative.

■ *Ron Paul? I have a hard time seeing him in a conservative light. He is such an avant-garde thinker.*

He is certainly distinct from the so-called conservatives, who have taken control of the Republican Party. That is how words get taken over by vested interests.

■ *I don't meet anyone anymore who is not concerned about his or her personal security financially. How do we speak to these people? We are talking about rather utopian theories and eventual possible realities—but how do we reach the person who has just been laid off—with no prospect of a job on the horizon. What do you say to people to help them find a direction to dig their way out?*

Those are certainly viable questions and basically what people have to do is to organize themselves in their own communities to collaborate and help each other through this transition period. As I said earlier, I believe we are going through a metamorphic change, like the change from the caterpillar to the butterfly, and if you look at behaviors of those two creatures, the caterpillar's purpose is to eat and grow. That's all it does; it eats, and it grows, getting bigger and bigger, and then it goes through a process called "molting," by which it sheds its skin and then develops a new skin. It goes through that process four or five times until it stops—but eventually it does stop eating. Its behavior can be devastating in that period, because it can completely devour plants in the process of growing—but it does eventually stop growing. There is something in its program that tells it to stop growing.

Then, it spins a cocoon and goes into a period where its body virtually disintegrates and the caterpillar turns into a new creature—a butterfly—during this metamorphic process inside the cocoon. What is going on is that the cells that form the butterfly were there, in the caterpillar, all along—but they were dormant through the caterpillar stage. But when the caterpillar goes into the cocoon, these cells become active and start organizing themselves into the organs that will form the butterfly body. At some point, when that process is completed, the cocoon will split open and the butterfly will emerge. The behavior of the butterfly is quite different than the caterpillar. It flies around, sipping nectar from the flowers, and pollinates them in the process, supporting their reproductive cycles—it has sex and continues its own reproductive cycle, so that the females will lay eggs, that then produce more caterpillars, and the cycle begins all over again.

If that is a metaphor for what we are going through today, then we should not look toward a restoration of the way things were. As far as the economy recovering, it is not recovering. We are moving to a steady state economy where production will level off and where we will supply the essential needs of people more efficiently, and we will stop producing a lot of waste and things that people don't want or need. There is a current debate in Congress whether to fund a new fighter aircraft. The Pentagon doesn't want it and yet Congress is pushing for it because it supports the economies of the districts in which these congresspeople are located.

We have to stop this kind of thing. We have to stop producing junk. The distance between the factory gate, where products emerge, and the landfill, where they go at the end of their lives, has become shorter and shorter over time. We have "disposable this" and "disposable that," and until now the emphasis has been on increasing labor productivity. We have to focus more on resource productivity, more value out of less stuff.

■ *Clearly, the way to do that is to dramatically reduce our consumption.*

Yes, to consume those things we really need. The problem is that to corporations, labor is a cost to them and they maximize their profits by reducing costs and increasing revenues. So they go to places where labor is desperate for work and income—that is basically what has happened in the U.S. and Europe over the last decades. Manufacturing has gone to Asia, where labor is abundant and cheap—so now the Western countries have very little in the way of manufacturing capacity left.

If we are going to recover economically, we have to focus on the localization of manufacturing—of producing more for local use, rather than for export or to distant markets. With the end of the cheap oil situation, we are going to see an increase of costs in transportation, which is going to favor local production for local use. With the financial collapse, we are seeing economies pull back in that direction as well. China, for example, has been greatly dependent on the U.S. market and they have accumulated a tremendous amount of U.S. dollars as a way of maintaining their sales into the U.S. market. But now they are looking at the prospect of those dollars becoming less and less valuable, so they're going to have to take the loss if the value of the dollar declines in relation to their own currency.

■ *Do they have to maintain a certain margin of dollars in their central monetary system, because they know if they revert out of dollars, the U.S. will pull back on trade?*

No. The more severe penalty will be they will force the value of the dollar down and that will make their exports to the U.S. market more expensive. One way or other they are going to have to reduce their emphasis on production for the export market and increase the emphasis on production for their domestic market.

■ *How does all of this relate to gold and silver? Many people believe or are instrumentalized into believing that they should amass gold as some sort of hedge or security against the financial disaster?*

That is a good question. Historically, when there have been financial disruptions or concerns over the value of political currencies, people have fled into precious metals. Gold and silver and other commodities and collectibles—anything that isn't dollar denominated or denominated in the currency of the country in which they are living—as a store of value, gold and silver may have a place in one's investment portfolio. However, as an exchange media, I don't think we are going to go back to gold and silver. We have these contradictory functions that money is supposed to serve.

In Economics 101, you learn that money serves the functions of mediums of exchange, store of value, and measure of value. Well, when gold and silver were the dominant forms of money, they did serve all those functions. But with credit money, that is not the case. Credit money is, or should be, purely an exchange medium. If you want to store value, you do that in different ways. You buy financial instruments like corporate stock or corporate bonds or CDs at the bank—or some other long-term financial instrument. Or you buy something like a physical commodity or real estate—those are mainly the options that people use to try to protect their nest eggs.

■ *They certainly are shaky at the moment. Of course, like the essence of all spiritual traditions, these global events are reinforcing the principle that nothing is permanent. If you look at the Buddhist teachings, the tradition teaches you to free yourself of material possessions, because you can't hold on to them. The desire for material gratification and the need to hold on to possessions all create suffering.*

We see this playing out in our financial decline—in real estate, which we saw once as the stronghold of any financial portfolio, we see values crashing and people even losing their homes. It is even written into laws in many countries that the government can take your property away anyway, in the name of "eminent domain." Nothing is secure in the sense that we have thought of it—through possession.

That's true. There is always the possibility of government coercion in dispossessing people of whatever property they happen to have.

■ *It seems impossible that the government can just come in, give you a few bucks for your home and property, and just say, "We're putting in a highway, or a commercial center—ciao!" This is allowed and tolerated all over the planet!*

Yes, there has been tremendous abuse by governments of the privilege of eminent domain. In 1933, when Franklin Roosevelt became president in the midst of the Great Depression, he called in the gold. He said, "It is now illegal for Americans to hold gold.... You must turn your gold into the government." And in return, they were given, I think, twenty-two dollars in paper currency for every ounce of gold. As soon as all the gold came in, then it was jacked up to thirty-five dollars per ounce.

■ *Do you foresee this happening again?*

It could very well happen. I don't really know.

■ *I had a friend years ago who was a collector of old gold coins—he had a huge quantity. He got a letter from the bank where he held the coins in a safe deposit vault, stating that hoarding the gold was no longer "allowed" and that he had to come to the bank and remove it.*

I asked him, "What does this mean? Are we going to have to turn our gold back into the government as we did back in the war?"

He said, "Yes. The government is saying that it isn't legal to 'hoard' gold. That's a precursor to the next step."

That's the kind of situation we have to anticipate. Recently, there was a private mint producing "Liberty Dollars." They were raided, and all their metal was confiscated by the government. They were also issuing certificates that were redeemable for gold and silver. That was in a private warehouse. And all of those metals were confiscated by the government. This was just a year or so ago.

■ *This is what I meant earlier when I asked you how these alternative economies can exist without the government coming in and saying, "Thank you very much," and stopping it. Of course they have every reason to stop these complementary currencies or the creation of these currency exchange systems.*

That's true—which is why people are going to have to organize themselves and assert their basic rights.

▓ *We've got a lot of work to do—that is for sure. It is so exciting to hear a person with your background and knowledge speaking with a truly spiritual and humanitarian slant on the fiscal situation, our personal security financially—something exactly timely and ultimately in the foreground of our experience, right now, in this time of transition.*

How do you foresee some of the new things that are evolving in our societies—such as the Web-based commerce that you refer to in your work? Can you give us an idea of where that is going?

Sure. You have things like eBay, Amazon.com, and all manner of online marketplaces that are growing by leaps and bounds and doing huge amounts of business. I personally buy books on Amazon and other things on eBay. As we generalize those online systems and include not only online marketplaces but also combine them with credit clearing systems, with a new standard of value or unit of account, then we can create Web-based exchange systems that are independent of the political currency systems as well.

▓ *Don't you think Big Brother is going to step in and slam his big boot on these kinds of schemes?*

It depends on whether or not they perceive that their time is coming to a close and that they are going to have to make some changes in order to have any kind of ordered society at all. This dominant paradigm is leading us to destruction and it ultimately cannot continue. What I am offering is solutions that benefit everyone. People are going to have to work out the details for themselves but it is going to be, I think, a more decentralized system of politics and economics.

Jane Jacobs, the economist, said that it is not national economies that are the salient entities. It is cities that are the salient economic entities. The national economy is simply an aggregation of city economies, so we have to look back toward the local level and see what needs to be done to support the health of communities. That is why I called my second book *New Money for Healthy Communities*, because this is the way we create healthy communities.

■ *And again, it is so timely. We have been living in such alienation from each other. In that almighty pursuit of runaway consumerism at all costs, we have moved further and further away from that community awareness. It is so timely (and I said earlier, perhaps that is why it must all come down) that we are called back to serve the community—back to our humanity—stirred into action at the brink of near disaster.*

Yes, and it depends on our recognizing the fact that we are not separate entities. There is a saying, "No man is an island," and we have become an isolated society. Most of us don't even know our next-door neighbors. We pay a price for that. This isolation is very costly for us. If we are going to solve our problems, we are going to have to come together and cooperate, taking advantage of our group intelligence and wisdom. There's a tremendous process of reorganization that has to take place, and that is what is going on around the globe.

■ *That is the wonderfully optimistic thread that is woven through your work: yes, there is disruption, but underneath this breakdown lies the incredible opportunity to reexamine the nature of human potential and our role in community, in the greater society, and, in the end, as the true caretakers of our entire world.*

There has been a big movement in recent decades toward intentional communities and toward community building. The Transition Towns movement, which is gaining tremendous impetus, is spreading all over the world. There are sustainability groups organizing—we have one in Tucson, called "Sustainable Tucson," which is an official Transition Towns organization. Basically, it brings people together to look at the problems we have and how we can move into the future in a sustainable fashion, because we are facing this multidimensional crisis that I mentioned: resource depletion, global warming, peak oil—all of these things involve the breakdown of institutions. They are indications that things have to change.

■ *Yes, they have to change urgently. What do you say to those who want to learn more—where can one seek out these organizations and find out about them?*

Transition Towns USA gives you a rundown of the various Transition Towns in the U.S., or anywhere. You can find this by searching the Web. From the business standpoint, there is the Business Alliance for Local Living Economy (BALLE) that involves small and medium-sized businesses around the United States, businesses that have a different kind of vision and motivation and are looking at how to compete with the "big box" stores and national chains, and how to move into a sustainable economy to produce what people need and to distribute it efficiently.

■ *Like you, Thomas, I see what is unfolding as the opportunity for immense change at a time when we know we are shifting away from the old and into a new paradigm.*

What advice would you give to those who are concerned, but still do not know what to do about the changes that lie before them?

I would advise people to get to know their neighbors, get involved in their communities, plant a garden in their backyard, take responsibility for their own health, and learn as much as they can about what is going on, so that they can be part of the change, rather than part of the problem.

■ *Indeed. Be the change you want to see in the world. Reading the works of Thomas Greco will help you rev up your engines and get into the positive mindset, through which you see what is breaking down as a necessary process (the dissolution of the cocoon) for the beauty that is about to emerge from the chrysalis.*

■ ■ ■ ■ ■

Thomas Greco's books are:
  Money and Debt: A Solution to the Global Crisis
  New Money for Healthy Communities
  Money: Understanding and Creating Alternatives to Legal Tender
  The End of Money and the Future of Civilization

Be sure to investigate his vast library of information by visiting:
  the Web site: www.reinventingmoney.com
  the blog: www.beyondmoney.net

Animal Activism

Mark Hawthorne

Mark Hawthorne gave up eating meat after an encounter with one of India's many cows in 1992, then went vegan a decade later, upon visiting Animal Place, an education center and sanctuary for farmed animals, near his home in Northern California. Beginning in 2004, Mark served as a contributing writer for Satya, until the magazine ceased publication in 2007, and he is a frequent contributor to *VegNews*. His writing has also appeared in *Vegan Voice, Herbivore, Hinduism Today,* Utne.com, and newspapers across the U.S. Mark shares his vegetable crisper with five rabbits rescued by SaveABunny, which works with San Francisco Bay Area animal shelters to find loving homes for rabbits. Much of Mark's writing can be viewed on firstprecept.com. He is compassionate, determined, and selfless in his quest to end the suffering and bring dignity back to the animals of the Earth.

The day of the interview with Mark, I rescued a puppy that had been dumped in the garbage bin, along with six others (who also were saved), by some utterly numb human being who figured it the easiest way to get rid of the problem of unwanted pups—knowing very well that they would be ground alive in the sanitation machine's grinders.

Leaving seven innocent baby pups to die, to be ground alive—is it possible that the human race has been reduced to this? I rescued him—that makes four dogs at the Cori household—I thought it was synchronistic with the fact that this interview was held the same day.

■ ■ ■ ■ ■

■ *To start, Mark, can you please describe that moment of conscious awareness that triggered you to become a vegetarian?*

Back in the '90s, I took a few years off to travel and I ended up in India, and I ended up living in the Himalayas with a Buddhist family (I was a meat eater at that time). Everything I ate then came out of their garden and I became, essentially, a de facto vegan. I just felt great, living at twelve thousand feet in the mountains. One day, a cow wandered into the garden, and we just looked at each other— kind of a mystical moment for me—and I realized that this being deserved to live just as much as any human does. I decided, then and there, to stop eating meat when I got back to the United States. I did, and over a period of time I transformed into a vegan.

■ *I'll share my story into vegetarianism as well. Still a meat eater about twenty-four years ago, I went to the butcher and complained that the steaks I had cooked the night before were extremely tough. The butcher looked at me vacuously and said that it depended upon how much the animal had suffered at the kill. He explained about the adrenaline rush at the moment of death and that the more the animal suffers, the more*

*terrified it is the moment of the slaughter, the tougher the meat will be ...
and that it wasn't his fault—or his problem.*

*In that moment of instant awakening, I asked myself, "How did you
never think of this in your entire life?"*

*I decided then and there that I would never eat meat again, and I
haven't.*

*It's pretty amazing when this veil of ignorance lifts! The next day, at
Eastertime, I was on the highway and there was a huge truck full of baby
lambs, on their way to slaughter. I looked at their little, sweet, innocent
faces staring out from the truck bed in terror and thought ... how hypo-
critical we are as a society. On the one hand, we give the children little
gifts of fluffy, stuffed animal lambs and bunnies as gifts at the Easter celebration, and on the other, we serve their little dead bodies up for our
feasts. Those eyes, filled with terror, still haunt me.*

Yes, there is an inconsistency in how we raise our children. On the
one hand, as you say, we give them cute stuffed animals and baby
ducks—they watch cartoons, where animals are the stars, and then
we feed them chicken nuggets and hot dogs and all these horrible
things. Some children make the connection and others don't.

Sometimes, later in life, they feel betrayed.... "How could my parents have done this to me?"

It's quite an awakening for them.

■ *This leads us to the question of the essence of life and consciousness in
animals, and how we perceive it. Many people don't really believe animals even have a soul! This is a question that seems to go on and on in the
theological halls of debate for eternity—and probably will continue to.*

Obviously, I definitely think that animals have souls. I can't look at
an animal and think it does not have a spirit and that it wouldn't
also have an afterlife—if I have one....

■ *I think you'll agree, not eating meat raises your sensitivity. I think processing meat through your system lowers your vibration and takes you out of synchronicity with other life forms. The more you eat meat, the further you move away from sensing the life force of other living beings and the conscious energy fields they emit.*

Absolutely. You're consuming a corpse: eating a dead thing.

■ *Right! My message is just that: You are seeking to activate the light body, refining the dense body, and yet you are eating dense, dead animal flesh! It is an utter contradiction in terms—and intention.*

Of course, it is important to try to educate people without preaching to them—an art I am sure you have mastered—so that you do not alienate them. People change when they are ready to change. We cannot present a case for the cause of animal protectionism, or vegetarianism for that matter, by presenting a case that is so overwhelming, they simply tune out.

I agree. It is important to find a balance in how we approach people. It is so important for animal activists, like us, to be friendly and approachable, but on the other hand we do have to speak the truth. Sometimes the truth is ugly.

■ *Especially at this moment on our planet ... there's so much un-truth, that standing up for any belief system that goes against the grain ruffles a lot of feathers (excuse the animal metaphor!) and that is what we're doing. My congratulations to you, for being such a leader in this very important quest of helping people wake up.*

Thank you. I look at myself as a conduit for animal activists around the world. I interviewed about one hundred twenty of them, putting their thoughts and their successes into book form, to inspire others. I really see myself as a vessel for all of their work.

■ *It is an amazing book, so full of wisdom. It is a compendium, of sorts, of information for people who want to really get involved in animal activism.*

Let's talk about what we are fighting for, here. Can you give us a general overview of the work you do as an animal activist? What are PETA and other major animal rights organizations really doing out there to make a difference?

Nonhuman animals are the most abused and exploited beings on the planet. That abuse includes: animal agro-business (factory farming), which is probably the primary one that people are concerned with; ocean animals, being pulled from the sea to become somebody's dinner; medical research; and product testing. In both these cases, animals are kept in tight cages and used to test a product, experimented on—often without any pain relief.

There's the fashion industry, which includes the use of wool, fur, and leather. Some people think of these as by-products of animal agro-business, but, in fact, in many cases, they're not. There's the entertainment industry, which includes circuses and zoos, for which animals are often trapped in the wild. If they're young, their parents are killed and then they are exported out of their environments, often taken to a foreign country, where they are held against their will—trained to perform tricks that are not natural to them, for human amusement. There are sports, such as dog and horse racing; hunting (which definitely doesn't seem like a sport to me!). There are "canned" hunts, where animals are held on a small piece of land and then they are let out, when the hunter approaches, so that the hunter is guaranteed a "kill." There's no sport in that! Even domestic animals are grossly abused.

These are some of the areas of abuse that we are fighting against.

▪ It is so hard, talking about these issues, not to get emotional and even depressed. You do discuss this in your book—how not to get burned out when you are dealing with so much abuse and suffering. How do you personally handle it?

I think that the activists who are the most successful—that is, the ones who stay in this work the longest—are those who manage to find a balance. We are just human, so we need to find a way to be active and dedicated, and at the same time, to protect ourselves. The formula that I developed, from my observations of other activists and for myself, I call the "A-C-T-I-V-E" approach. Each letter stands for one of the steps:

- **A:** Allow yourself to be human. We're not superheroes—we need to take a break from it once in a while, vacation, laugh with friends, have a good time, and try not to feel guilty about taking the time off.

- **C:** Create something tangible to remind you of your victories: that could be a file with your letters to editors; a scrapbook of a campaign you worked on; a Web site—anything that reminds you that you are being successful in your fight.

- **T:** Talk to someone you trust. This could be someone inside or outside the movement, family, friends, partners. If you have just one person with whom you dare to be yourself, that is a true gift. Animal activism is an emotionally loaded endeavor, so it is important for us, as sensitive people, to unburden ourselves.

- **I:** Ignore upsetting text and images, especially if you are feeling disempowered or discouraged, feeling bad about something you've read or seen. Put that aside for a while—don't watch that horrible video.

- **V:** My favorite—visit an animal sanctuary. We have a lot of them in the United States and I know there are many as well in Australia and in the United Kingdom. I don't know about other countries. These are centers where animals from research centers and factory farms are brought to a home, a piece of land, where they're allowed to live out their lives. You can volunteer at these places. I do volunteer at an animal sanctuary here in California, and I am amazed at how many activists have never rubbed a pig's belly or whistled to a turkey to hear him gobble back ... or watched a hen take a dust bath. You know, hens in their natural environment love to take dust baths, as it helps them get rid of parasites. All these things that animals can do in nature (if they are allowed to be free and not confined in these giant animal factory farms), activists can experience at these sanctuaries.

- **E:** Exercise. I think it is really important that we get out and walk, run, hike, swim, go to the gym, practice yoga, meditate ... whatever we can do for our bodies and minds to help release a lot of the heavy energy from our work.

I think that these six things provide a great formula for activists, to avoid burnout from the work of animal protection and activism.

▧ *It is important to point out that it is not necessary to devote your entire life to this cause—and that even if you dedicate a very small portion of your time to helping animals, it is still very valid and so significant. I believe it is important that we all give what time and effort we can....*

Yes, you surely don't have to be me! I dedicate about twenty hours a week to this work.

■ *Even an hour a month will make a difference.*

Certainly! If everyone on the planet dedicated an hour a month to the cause of animal protectionism, the world would be incredibly different.

■ *I interviewed author Penelope Smith, who has contact with animal spirits that have passed over to the other side. I think she has done a lot to sensitize people to the fact that animals have souls and, like us, they have an afterlife ... and that we need to be aware of that connection, on all levels.*

I agree, wholeheartedly.

■ *So, what do you recommend to people who want to become more active in working with shelters and other aspects of animal protection? How do they make those connections—do you have listings on your site?*

Yes, there are several links on the site and these are also listed in the book. You can also do a Web search, specific to where you live. Do the research. I am also available to answer questions sent to me through the site.

■ *Reverting back to the question of vegetarianism ... I would like to also discuss the case against eating meat, at a merely practical level. With all the diseases now being transmitted from animal meats, not to mention the hormones, antibiotics, and pesticides in their bodies, it is a case for people who may not even be sensitive to the question of animal protectionism to consider what illnesses can be picked up from the meat they are eating.*

Well, yes, inevitably this is one of the main arguments for becoming a vegetarian. Besides (as we have been discussing) not eating animals because you do not want to contribute to cruelty, and not wanting to contribute to the greenhouse gases being produced in these huge agro-businesses, there is of course the question of your own health. Eating meat can lead to cancer, heart disease, stroke, obesity. There are a number of serious health conditions a human being can acquire from eating meat.

There's a debate as to whether human beings were ever meant to eat meat and I try not to get into that argument, because I don't think it is that important.

■ *I agree with you. It is an endless tunnel. Once you start getting into it, people can become defensive, or cite arguments such as "the food shortage" (which I don't really believe exists). Most meat eaters simply have a lack of fantasy and creativity about the alternatives to meat diets.*

Right. We put so many resources into raising livestock. We clear-cut our forests to raise grain to feed them. It takes sixteen pounds of grain to produce one pound of beef. Those sixteen pounds of grain, going to one animal (just to produce one pound of beef), could be fed to human mouths, so that argument simply doesn't hold water for me. There are many similar statistics regarding water resources, goods, and other energy resources needed.

In late 2006, the UN released a report that showed that animal agro-business was responsible for more global warming than all the transportation in the world combined! All the planes, ships, cars, trucks combined contribute less than all the animal agro-business does to global warming.

■ *Let's be sure everyone understands exactly what you're talking about when you refer to "animal agro-business."*

Essentially, I am talking about the business of raising animals for food. Some of these definitions could be factory farming or what I call "CAFOs" (Confined Animal Feeding Operations). They've moved animals from the small farms (that many of us are familiar with from decades ago) to these giant factories.

They've moved them indoors (where they never see natural light), because they found that by giving animals less room to move around, they ate less feed, and they produced more meat ... or more eggs, or more milk. As they realized this was the case, factory farming became the norm. Old MacDonald's farm is the myth. There are very few small family farms, now, that are really effective. Now it's all these giant, multinational corporations that are producing the "animal products."

■ *Do you want to give people the scary information, about how small the cages that animals are confined to really are?*

I have been inside a battery cage during a rescue for egg-laying hens. Basically, they have about sixty-seven square inches per bird. A hen needs at least seventy-two inches, just to spread her wings. These cages are about the size of a file cabinet drawer. There are six to eight birds in one cage!

These birds are denied all their natural instincts. The tips of their beaks are cut off, without any pain relief, so that they don't hurt each other. They stand on wire. They cannot spread their wings. They cannot nest, so there's no place to lay their eggs naturally. In many cases, they are stacked in tiers, two to four high, so the birds on the top level defecate and urinate on the ones below. That's how these animals live: in darkness, breathing ammonia, for between one to two years....

■ *Why ammonia?*

Because there are manure pits below, and, as the feces disintegrate, they release ammonia—you find this in factory farms, where they are raising meat chickens as well—they call them "broiler" chickens. You walk into these places and you are just assaulted by the ammonia; it burns your eyes ... and these animals are breathing it their whole life spans. So they fill them up with antibiotics.

■ *And therefore people are ingesting that as well.*

That's right.

■ *It is so hard to deal with this incredible abuse and inhumanity. I commend you for it—and all the activists, like you, who are doing this intensive work. This is reality. People need to look at it and listen in order to make changes. We, the human race, are supposedly the most intelligent species on Earth and yet this inhumanity continues.*

I ask people to be strong and to be willing to hear the truth, so that they can do something about it.

Yes, it is difficult. I've been inside factory farms; I've done rescues. I've worked with animals that have been abused. I have five rabbits at home who are "special needs" rabbits, because they were abused and neglected in some pretty horrible ways. Yet, living with these animals, I see so much hope. I see how easily they can come to trust humans again, and I find a lot of encouragement in that. If we work together, if we are compassionate enough to spend even an hour a month doing something about it, we **can** make a difference.

Agro-business and the entertainment and fashion industries ... the sporting industry ... they're all nervous about animal activism. They don't like what we do, because they know we are effective.

■ *In fact, in the book you speak at length about how to deal with persecution. You talk about the police, explaining how to deal with "reactive forces" against the work. I would love for you to share some of these insights with people, so that they know how to deal with what you're up against.*

I include an appendix in my book that deals with the civil rights codes in some countries around the world in which the book would be readily available. Basically, I think the best advice is—if you encounter a law enforcement official, while you are in the act of animal activism that you believe to be legal, you don't need to say anything. Get yourself an attorney. Don't admit any guilt . . . because, as far as you're concerned, you haven't committed any crime.

■ *You're talking about a situation where one has actually gone in and done a rescue?*

No, I'm talking about a situation where you might be in a public place, for example, and you're distributing information about animal rights or veganism . . . something of that nature. Maybe a store owner nearby is upset and calls the police. The police come and talk to you and, in some cases, you might be arrested. I don't know—it depends on the country and circumstances. I'm not an expert on civil rights laws.

■ *Well, in the U.S. today, would you be arrested?*

Probably not. I don't think you would. I think you would have to be doing something destructive, in order for the police to have cause to arrest you. I think my advice holds true for that instance as well. If you find yourself arrested . . . you need to get a lawyer.

I don't want to dissuade and discourage people, or frighten them, to think that is what is going to happen to them.

■ *I agree with you. But on the other hand, people need to know how hard you are all working and the kind of personal danger and risk you face in doing what you do to help the animals. You do this willingly, proudly, convinced ... because you are committed. It is important for people to know you do not take this lightly and that you are willing to face even persecution. They do need to understand that getting behind organizations like PETA and the many more that you name in the book means supporting people who are risking even their lives: like the Greenpeace warriors, who dangle from ropes off ships. You can support them by donations, by being aware of what peril people place themselves in in pursuit of their beliefs—honoring them, spreading the word.*

I think this is empowering rather than frightening.

There are indeed incredible people out there, like Captain Paul Watson, of the Sea Shepherd Conservation Society, who actively goes out and chases down Japanese whaling ships. He's been arrested any number of times: he's been in jail; he's been beaten; he's been shot at.

He's one of my heroes—out there doing the good work. He's just a complete activist and he's a great example of what one committed person can do for animals in the world.

■ *He is one of my heroes as well. Do you have any information as to the situation now, with regard to whaling in Japan?*

I'm not sure exactly what the situation is. I know that, at their last whaling expeditions, the Japanese ships came up short. I think they got about half of the whales they had hoped to kill, and I think that is in no small part due to animal activists.

■ *I wonder if it is because the whales are beaching themselves, so resigned to human abuse. I believe the ELF (extremely low frequency) underwater sonar waves, being blasted by the military, are doing immense damage to the whales and the dolphins, driving them to suicide.*

Yes, I agree. Unfortunately, our government in the United States is responsible for that too.

■ *Is there any active campaign against the use of these sonar waves?*

That is a great example of outreach that people can get involved in. You can learn about what your own military is doing or what your own government is doing. In the United States, for example, the USDA has an offshoot called "Wildlife Services." This is a part of the U.S. government that goes out and traps and kills animals on public land. They generally do it because these animals are either foraging for livestock or they're just considered "nuisance" animals, like a coyote, for example, which goes after sheep and goats . . . and other livestock.

Wildlife Services will also kill a wild animal for eating someone's flowers, or for frightening somebody. I don't think that most people know that their tax dollars are being used for that. That's just one example of something to be upset about that your government (at least the U.S. government) is doing.

You can definitely talk to policy makers; you can talk to your elected officials. Arming yourself with information is important; you need to be aware first of all of what is going on. Going to a Web site like PETA's, or another organization's site in your native country, is important. So is learning about the issues at hand locally, and then contacting policy makers, whether that is an elected official, or whether it is a company that is doing animal testing or abusing animals in other ways—to tell them that you want to see it stopped.

Whether because it's a potential lost vote or whether it means lost revenue, your voice counts. It's important to let people who are in the position to make decisions know how you feel.

■ *I do also believe that it is important that people really understand that their voice—and that one call or letter—can truly make a difference. So often people feel so powerless to effect change that they don't make that call or sign that petition.*

The truth is that most people don't write letters. Most people don't make phone calls. Editors know this; elected officials know this. So, when a person takes the time to write a letter or send an e-mail or pick up the phone and talk to a policy maker, they realize that that person speaks for a larger group . . . the group that doesn't pick up the phone, doesn't write or send that e-mail.

You might think that if you send a letter to an editor and it doesn't get published then you've wasted your time—but that's not true. You are educating that editor and you are letting the staff know that there's a group of people who are concerned and interested about this particular topic. The same is true with elected officials. They know that most people aren't going to write to them or call, so if you do, they look at that as a potential vote. They say, "This is somebody who cares enough to take the time to do this—I'm going to pay attention to it!"

■ *Do you really think so, considering the lobbyists behind them, pushing their agendas?*

Yes, I do think so. I realize we have so many lobbyists pushing through their agendas, like the Farm Bill—but yes, it makes a difference.

■ *What do you think about the relatively new phenomenon of cloning animals? What do you feel this means to people who are going to be ingesting this meat?*

I believe that exploiting animals in any form is wrong. As I have said, I don't think we should be using animals for food, for clothing, for testing, for entertainment ... none of this. So, I feel equally strongly about cloning animals, because it is exploitation and unnatural.

As far as the safety of eating this food is concerned, I don't believe there is any proof that this is in any way safe to eat. I know that some agencies have declared that it's safe to eat, but how do we know? How do we know that, ten years down the line, you're not going to acquire some new disease, some new strain of cancer? They keep telling us that the food supply is safe, and yet we constantly have issues with new diseases, such as mad cow, bird flu, swine flu, and other issues that are involved with the illnesses animals acquire. I don't think anyone can say with certainty that cloned meat from animals is safe to eat.

■ *And then, the very associations that are reporting it to be safe are probably "unsafe" for us as well!*

That's right! They have a definite monetary interest in what we eat.

■ *We've certainly seen in the last ten years, particularly, that the food supply is anything but safe! Particularly harmful, of course, are the animal "products."*

Right. And whether or not they are determined to be "safe," we do still agree that eating them, overall, is bad for your health. For example, I don't think that having rotting beef in your colon is good for you. Neither is milk. We are the only species on the planet that drinks the milk of another species and that drinks it after we've been weaned! Does that seem natural? No, it certainly is not. I think that the suffering that animals endure in the dairy industry and the egg industry are worse than anything that happens in the raising of beef.

Just to speak about dairy cows for a moment: these animals are artificially inseminated every year, so that they lactate. They are kept pregnant, so that they give milk, and their babies are taken away—a day or less after birth. The newborn females are recycled into the dairy industry—the males either become veal or they become beef products. They may not be considered good meat animals, but they are definitely killed.

The female cows providing the perennial milk last about five years, until their bodies are exhausted. Then, they're sent off for slaughter. Their bodies are so depleted by then that they basically can't be used for anything other than hamburger. They're just not worth anything more than that.

In nature, a cow will live about twenty to twenty-five years, so you're talking about one-fifth of a normal life span that she has spent, imprisoned, in a factory farm, on a milk machine. All her babies are taken away. The separation between a mother and baby cow is just unbelievable. It's heart wrenching to hear the cries from the two animals, at that moment of separation.

When I learned about what goes on in the dairy and egg industries, I determined that I had to go vegan. I just couldn't support them anymore.

■ *It is so important that you do what you can and that you progress as best you can.*

Yes, and it is important to remember that every animal-based product has a substitute that you can replace it with. There are great soya products, vegan ice cream, baking recipes that don't call for eggs ... tofu. There are plenty of alternatives. When I learned what happens in the egg industry, I couldn't eat them anymore.

■ *How do you feel about eating plant foods? They too are conscious. Naturally, people ask vegetarians, like us: "If you don't want to eat meat—then why do you eat plants? They have feelings too!"*

And there have been some pretty remarkable experiments that prove this—like the work done by Marcel Vogel of the Stanford Research Institute. He attached an electroencephalogram to various plants and studied their reactions to various stimuli performed in tests, with pretty amazing results.

I have a two-part response to that question, when it is asked of me. The first is that plants do not have nerve endings—they don't have a central nervous system. They don't have a mechanism designed to help them escape pain stimuli. Therefore, I do not think they feel pain. The second is this: if you believe plants feel pain, then being vegetarian is a better resource for you, because the biggest consumer of plant life in the world is livestock. We raise crop to feed to livestock.

■ *Touché! That is a great response. I encourage people to be aware that all life is consciousness and to practice blessing the food before ingesting it, just as the native peoples have always done—honoring it. We do need to eat. We need to do the best we can to do as little damage as possible through our habits.*

Agreed. We need to realize that everything we consume, even if you are vegan, at some level harms animals or insects. We need to try not to be sanctimonious about it—animals don't need our purity— so it is important not to let the ideal of perfectionism get in the way of being an advocate.

■ *What is your next book going to be about?*

I am currently working on something that is intended to give activists some insights into some of the lesser-known abuses and crimes being perpetrated against animals, by the fur industry and the others I have mentioned. Some of the behind-the-scene things that the governments are doing, that we're not aware of ... and I'm speaking worldwide.

■ *That's very courageous of you, Mark.*

Yeah, I guess it is! I suppose it could get me in trouble!

■ *Well, as they say, "A man's gotta do what a man's gotta do!" That's what having a voice is all about. The mere idea of becoming an activist implies that courage will be needed. I do own that in my own skin.*

I just don't think there is any way else for me to live. I have to do this for these beings, who can't speak for themselves. I need to do it for them.

■ *Mark, you have a list of eleven things you can do today to help the animals. Can we share them here?*

Absolutely. They are:

1. Go vegan: If you can't, try giving up meat, eggs, and dairy for one day a week. See how you feel. Then try two days ... then three, and so on.

2. Set up an account on a site like iGive.com or goodsearch.com. These sites allow you to support animal groups with a percentage of your online shopping and I would like to stress that it doesn't cost you anything to do this. These are search engines that allow you to do your online shopping but they give a percentage to the organization that you select.

3. Contact a local animal shelter or sanctuary and ask about volunteering—we discussed that earlier.

■ And I'd like to throw in a little comment here: if you take home a domestic animal, take it home for life, not because the kids want a pet they saw on TV, or because that breed is a fad at the moment. These are not playthings. If you bring an animal into your family, make sure you are committed, and that this being doesn't become old hat and then abandoned, please! Having just rescued a beautiful little puppy from the garbage grinders, I feel particularly emotional about this.

Yes, I want to add something to that as well. I had a friend who worked at an animal shelter, who told me about a woman who brought in a cat that she had bought from a pet store. I kid you not: she brought it to the shelter because "it didn't match the curtains."

■ You can't be serious, Mark—this pushes me beyond credulity.

Unfortunately, I am. So I totally agree. If you bring an animal into your home, he or she is part of your family. Please care for them.

■ And they give you unconditional love. Just give it back—for the lifetime of the animal. How many people dump their faithful animals on the side of the road at vacation time? How many are left behind, with no idea of why they have been abandoned—and with no way to survive on their own?

Rabbits are another example of that. Every year, at Easter, people buy rabbits for their kids. Then, when the novelty has worn off, they dump them in a park or in a shelter. These animals are treated as toys—commodities.

■ When I brought the little pup (whom I named "Lucky") to the vet this morning, she sighed and said, "You can't imagine ... people stick six or seven puppies in a bag and tie it to our fence...."
    Alas ... let's get back to your list.

4. Order leaflets from PETA or Vegan Outreach, and begin leafleting when they arrive! The whole first chapter of my book is about leafleting.

5. Ask your school cafeteria or one of your favorite restaurants to carry vegan entrées.

6. Write a letter to the editor of your local paper.

7. Visit your local bookstore and pick out a book on animal rights.

8. Go to a vegetarian site by searching google.com—get a vegetarian starter kit and give it to a meat-eating friend or family.

9. If you have companion animals who are not spayed or neutered, make an appointment with your vet to have this done. (Note: a lot of organizations do this for free, so if money is an issue, you can investigate this as well.)

10. Order buttons and shirts with anti-cruelty messages that you get out to like-minded people.

11. Include a signature line in your e-mails, with links to one or two of your favorite animal rights videos or campaign. Let people know where they can go for information. Helping animals is a very easy thing to do.

■ *All great suggestions—thanks, Mark.*

I want to remind people that, as we travel our path of personal and spiritual development, aspiring to a better world, we need to look even at these dark happenings around us. **The buck stops here.** *Ignoring this pressing issue is not going to make it go away. We have to do something about this crisis and the cruelty. We have to support people like Mark and others working in the field, often risking their lives to make change. We have to teach the children—they are the future.*

United, we can *make the world a kinder, gentler place for the animal beings that have every right to share this space in dignity, freedom, and love. Even an hour a month can make a difference.*

▪ ▪ ▪ ▪ ▪

For more information about Mark's work and the resources he makes available, see: www.strikingattheroots.com.

The Hour of the Dolphins
and Whales

■ ■ ■ ■ ■

photo by Marji Beach

Hardy Jones

Hardy Jones is a former journalist with CBS News and UPI. He has been making television documentaries about the oceans and marine mammals for more than twenty years. "The experience of forcing Japanese fishermen to release hundreds of dolphins, simply by pointing a camera at them, led to the original concept of BlueVoice.org," he says. "The advent of the Internet has given us a tool of unprecedented power to end some of the brutalities committed against marine mammals and the oceans." He has dedicated his life to saving them....

■ ■ ■ ■ ■

■ *Not too long ago, I saw a painful and tragic film about how the Japanese are slaughtering dolphins—a very difficult thing to watch. It was so compelling, I immediately went to my computer and searched the producer, Hardy Jones, who replied immediately, openheartedly, and agreed to give me this interview.*

Japan continues to take whales under the guise of "research." And now Denmark has decided that we need to kill more whales. Apparently, it is not enough that they're dying from human neglect, our toxic pollution, overfishing, destruction of their habitats. Denmark has petitioned the International Whaling Commission to allow Greenland to slaughter fifty more humpback whales.

That must not happen.

What an appropriate setting for me to be talking with this guardian of the seas, Hardy Jones. I am here in Newfoundland, on a whale study expedition, getting so close to these majestic beings that I have been almost able to touch them. And that is no easy feat, out on the choppy, deep waters of the Atlantic.

Hi, Patricia, you're certainly in a wonderful place to be seeking out the whales—Newfoundland.

■ *It is! I am here on a whale study program and I have been blessed with seeing lots of dolphins, some sperm whales, and humpbacks. Apparently they feed on capelin, a small fish that spawns in these waters. So they're coming in now to feed. How is that for perfect timing and location for this conversation with you?*

It's actually all the more relevant, because the whales that you are looking out on are part of the population of the ones that could be traveling up to Greenland and then will be under the gun. The humpbacks of the western Atlantic winter down in West Indies, around Puerto Rico and islands like Saint Lucia and Dominica, all the way down to the coast of Venezuela. Then, in the summertime, when they need to feed, they migrate up along the eastern seaboard of the United States. They're quite visible off the coast of Massachusetts on Stellwagen Bank, and then further off of Maine, and then up off Newfoundland. Part of this population then goes up to Greenland, so these whales that you are observing may be the ones that become targets, if we don't stop this.

■ *You went to the International Whaling Commission conference weeks ago and you reported to me that there was an official petition from Denmark that Greenland be allowed to cull fifty beautiful, majestic humpbacks. Why is this being considered? Why is Denmark making this ridiculous request and what is the status now of this situation?*

The request from Denmark to allow Greenland to kill the whales is a very curious thing. The aboriginal hunters of Greenland already have a quota to take minke and fin whales. They get a certain amount of minkes from that, part of which goes to the aborigines and part of which is sold commercially. The request that they take the humpbacks had, within it, the idea of diminishing the take of minkes and fins, so that the net take that would accrue to the hunters would, theoretically, be less when you add the humpbacks: less, in total, than their previous quota of minke and fin whales.

You look behind that and ask, "Why would they do that?"

The answer is pretty clear. The Japanese have come in and incited them to do this, through what you could call "bribery" or inducements of various kinds. Why would the Japanese want Greenland to hunt humpbacks? Well, the Japanese tried to get a quota to hunt humpbacks in the Arctic: fifty, by coincidence. There was such a worldwide outcry, because humpbacks are so loved, that they withdrew their request. So, now, what it seems they are doing is trying to get in the back door—to get somebody else to approve it ... and then they too will be able to go ahead and hunt humpbacks.

■ *Why must we kill whales? Can't we improve, as a species, and stop killing these magnificent beings? I just can't get my head around this. You did tell me earlier that the market value of one average-sized whale is around a half a million dollars, so that is an obvious answer. But surely there are other ways for the greedy to make those millions. When is human consciousness going to rise to the point that we stop allowing this brutal murder?*

I wish I had answers to those questions. There are actually two distinct reasons why the whales are hunted by Japan and by the aboriginal people. Those of Greenland do actually subsist on dolphins, such as belugas, seals, and whales, and one can make a case that they need that. For them to get meat flown in for them is terribly expensive. However, it has been amply proven that eating this diet, which once may have been healthy, is today highly toxic to those who consume it.

In other words, the Greenland aboriginals (it applies to Native Americans in Alaska as well, and the Inuit in Canada) are hunting in seas that are contaminated today with heavy metals and organic chlorines—two classes of toxic chemicals that are accumulated in these species they are hunting. One result is that, in Greenland, the very ratio of sexes born in the human populations has changed radically.

Down in the southern part of Greenland, the number of girls born to boys is now two to one. It used to be slightly in favor of boys. In the northern area of Greenland, beyond Thule, there are villages where there are **no boys born.** The obvious reason for this (the government of Denmark has even said so) is that the chemicals that have been absorbed into the bodies of the whales and the dolphins are basically estrogen imitators. Essentially, eating that meat is like taking the female hormone, estrogen.

■ *In fact, this is happening worldwide in human populations. This has everything to do with the reduction of sperm counts in the male populations, which has been reported as having dropped fifty percent in the last forty years or so. The estrogens we are dumping into the seas through our soaps, as well, and other contaminants we are dumping into the sea, are directly affecting human genetics.*

Yes, and in Greenland it is dramatic. How perverse do we have to get before we realize that first, you should not be eating these animals for your own health's sake and second, that we have allowed our oceans to become so contaminated that they produce this result?

■ *Let me add that if people are not going to protest such violent crime as that which we are perpetrating on the mammals of the sea at the ethical level, then, at least, your campaign to help people understand how the toxicity of eating these animals is so lethal, so toxic to the body—perhaps this will be a trigger for people to at least slow down on the consumption of these great sea mammals. Especially the Japanese: I do have to point a finger there.*

We have gone into villages in Japan and tested the toxicity levels of people—we've tested the dolphin meat in the markets. In the case of one man, we found eighty-nine times the allowable level of mercury in the body: eighty-nine parts per million, which is astronomical! We found many other people who ate dolphin meat, who had similar, extremely high levels of mercury.

■ *In fact, here in Newfoundland, I was told by a marine biologist that there was a fad not long ago of a strict "tuna only" weight-loss diet. Some people got deathly sick and almost died. One woman had such a toxic overload that, within a very few days of beginning this diet, she had to be hospitalized for mercury poisoning.*

That's not surprising. Tuna is at the top of the food chain. So are dolphins; so are humans. These chemicals bioaccumulate up the food chain. Even if a chemical company dumps a certain amount of toxin into the water, it counts on the fact that it will disperse. It does, in fact. In the ocean, it reaches down to very deep levels. What they don't count on, though, is that those chemicals, even in small amounts, then get into plankton—plankton is eaten by larger creatures, and so on, up the food chain. Each time a fish is eaten by a larger fish, that concentration is multiplied by a factor of ten. By the time you get up to the large predators like tuna, they're loaded with mercury and very dangerous to consume, in any quantity at all.

A pure tuna diet is not recommended!

■ *Neither is a dolphin diet, let us remind the Japanese! I heard that the Japanese are also putting whale meat into canned dog food. This really drives me out of my mind. Let's pamper our domestic animals with the precious bodies of our great whales, but let's not love the whales enough to protect them from such ridiculous abuse and waste of their precious lives.*

The Japanese government subsidizes whaling—it is not a profit-making proposition. The government gives subsidies in various forms to the whalers, so that they can continue to carry out this bloody business.

A lot of people scratch their heads and say, "Why do they do it? Why don't they give it up? They're losing money—they're losing the market for it...."

In fact, the price of whale meat has fallen. I think they have about forty thousand tons of whale meat in refrigerators that they can't sell ... so why do they continue with this?

■ *And why aren't those forty thousand tons of whale swimming around in the oceans, creating their beautiful music (we know they send their musical harmonies through the oceans), instead of rotting in a refrigerator?*

After many years of taking films for PBS, Discovery Channel, and international organizations, I thought, at the end of the day, that enough embarrassment and enough coverage of this (these terrible pictures on television) would cause them to pull back and stop the killing. It actually has worked, in many villages—there used to be quite a few villages that were hunting dolphins, when I started going to Japan, thirty years ago. Today, there is really only one village in the main island of Japan that still carries this out.

■ *What kind of a "hunt" is this? They round up these dolphins, who usually love humans, and are curious about us—they round them up and then brutally slaughter them, the bay filled with so much blood of these innocent beings that the water turns bright red. It is the quintessential bloodbath. We can't call that hunting—it's mass murder.*

Yes, that's what we're trying to stop. The good news is that it is down to one village now. They're the most stubborn. They claim that their culture and livelihood depend on it. There are really only thirty-seven guys who get anything from this at all. I was saying that I thought the exposure of this would stop it, but it is clear that this has not been effective, in this one village. Now what we are doing is to conduct these tests on the toxicity levels of the people who eat the dolphin meat, and I think that that is going to either force the government to shut it down, for health reasons, or it is simply going to cause people to say, "I'm not going to eat this."

■ I think that is a really brilliant approach, Hardy. It is positive and proactive, getting deeper into the human psyche and helping people to understand how, personally, it is destructive to their personal health and well-being. Otherwise, they just don't seem to find the motivation to stop it.

... And the Japanese themselves—the Japanese press and the medical people—are now taking up this issue. So it is not only coming from overseas; it's now coming from within Japan, which is a very positive sign.

■ So, what is your campaign doing to help stop the Whaling Commission from allowing Greenland to kill fifty humpback whales?

The first thing we need to do is to get the Obama administration up to speed on this issue. Admittedly, it is facing a huge number of challenges coming in, with this economy and with American troops fighting in Iraq and Afghanistan. But at the IWC, the American delegation was headed by a Bush appointee. The Obama administration had a very small presence there, so we have to let people in the White House know that millions of people care about the whales, and that they need to get it onto their agenda. That's what we're going to be doing over the next months—informing people about this impending catastrophe, and getting people cranked up to influence the United States government, because the U.S. has a tremendous amount of power in the IWC.

If President Obama, who has enormous popularity around the world, gets on the side of the whales, I think that will be one way of quashing this.

■ Well, he claims to be the "ecology" president, so, if his claims are true— this should be an issue that makes sense to him—something he can connect with.

Right, so it's a matter of getting people to write to the White House to tell President Obama that whales are very important to us. They are the symbols of the sea and they are highly threatened, not only by whalers, but by the toxins in the oceans. That becomes a great focal point: for us to save the whales and save the oceans. That is our message. We put out newsletters every month, giving people indications of how they can help and who they can write to—utilizing their networking tools for an issue like this. We can mobilize tens of thousands, even millions, of people, to make their voices heard.

■ *Now is the time to transform our concern, and our suffering about this issue, into real action. Never before has there been such urgency to get out of our boxes and take action to help. Aside from the fact that the oceans are toxic and that is creating toxicity in human beings, we simply can't sit back and let the oceans be stripped of the cetaceans—the magnificent varieties of whales and dolphins. We can't let this continue.*

I have called on people to send e-mails and letters to the prime minister of Denmark, protesting his actions and demanding that he retreat from this unthinkable course: this call for the killing of fifty more humpbacks by the Greenlanders.

Yes, that is also very important—letting the Danes know that we are not going to sit back and allow this to happen. The killing of humpbacks is, to me, just so perverse—they are so beloved to people. I don't mean that the minke and fin whales bleed any less or suffer less—but they are not as charismatic....

■ *They don't seem to be as playful, in fact. Here I am in Newfoundland: not my first time out with whales. Yesterday we were out on a Zodiac boat, which is pretty small and low on the water. Some humpbacks were out on the horizon and we came across a humpback with her calf. We turned off our motor and started singing, and they approached.*

419

The female came up ever so close to the boat, and then the calf came right up to us and even went under the boat. She allowed her baby to go under the boat and explore the human beings, who were making music— trying to communicate with them! This is an experience that just moves you to incredible heights, knowing that we do absolutely have interspecies communication with the great whales.

Unfortunately, the curiosity and friendliness exhibited by humpback whales makes it easy to round them up—and then they massively slaughter them. They have helicopters to track them; they have electronic equipment ... it's a new war they are waging.

■ *We've been sending a lot of e-mails to the prime minister, a lot of messages protesting his action to call for and encourage the slaughter of the humpback whales. One of my readers copied me on the message that he sent that said:*

> *Have you ever even been near enough to the ocean to see a whale in all its glory? Have you ever been close enough to look into its eye? If you had, you could not do what you are doing. We protest you; we protest your country, a country we believed was civilized and developed. We will boycott you; we will not rest.*

Boycotting the country touristically can help—so please do whatever you can do to help him understand that the world is watching.

Where does this stand now, with the International Whaling Commission, Hardy? What was the outcome at the Congress?

Well, it caused so much consternation in the commission that the chairman decided to put off a decision about it. They are going to have more meetings, where they will have to reassemble all of these dockets from all over the world again, and they will decide whether they are going to allow them to go in and hunt the humpbacks. So we have plenty of time to organize.

I just want to reemphasize the fact that, if a quota is allowed to Greenland, then the Japanese will have a case to say, "We have given this to these other people; give us the right too."

■ *Of course, it will trigger all kinds of new adventures into slaughter.*

Yes, it is a domino effect. When one country is allowed, then it is hard to make a case that another country shouldn't be. There was the issue this year about Japanese coastal whaling. If the Japanese got their quota to go whaling along their coasts, then the Koreans said that they would start. This is why we really have to resist, very forcefully, every effort to open up whaling of any kind.

I was encouraged that last Sunday [July 5, 2009], the *New York Times* wrote, on its editorial page: "There should be no whaling of any kind ... for any reason ... anywhere in the world."

That highly respected newspaper has put that message out. They also have an article in the Sunday *Times Magazine* that talks about communication between whales and human beings, so somebody at the *Times* has awakened to whales!

■ *That is good to hear. We can help stimulate the media, as well. What is your organization, BlueVoice, doing now, besides this important campaign to stop Denmark and Greenland? What other driving activities have you got going now?*

One of the things I wanted to mention, besides our campaigning against these atrocious practices, is that we go out into the ocean with dolphins. We have conducted some experiments with communication between humans and dolphins, with a Macintosh, which is in an underwater case. We have known a school of dolphins in the Bahamas for thirty-one years.

I first went out in 1978, before anybody had really spent any time in the water with dolphins. I have found a school of spotted dolphins in the Bahamas that are exceptionally friendly and I have maintained a relationship with them for this entire period of time. I have known some of these individual dolphins over this period of thirty-one years. We are making an effort, through our films and the Internet, to help people know just how magnificent these animals are—how friendly, intelligent, social, and curious they are.

■ *And as you said before—that's what gets them into trouble. Curious as they are, they often come up close to the ships of the so-called "hunters" (I prefer to call them "butchers"), who then kill them.*

Fortunately, the number of people who want to kill them is a very small number, but they can do incredible damage, especially when the humpbacks are so endangered.

If anyone wants to see any of our videos, they can come to our site, www.bluevoice.org; they can click on "Web Films" at the top of the page. One is narrated by Matt Damon, who describes our work. We had an encounter with a friendly sperm whale, off the coast of Japan, as a matter of fact, which just swam right up to our boats and then lolled around for forty-five minutes.

People can forward these links to their friends—that is another powerful way of working. If you look at some of this footage, and you see the mating rituals of these humpback whales, or you see a sperm whale swimming toward us ... you can get a visceral contact with the animals that are threatened, and which the Japanese and the Greenlanders are hoping to kill. People can get active, by sending these links to their friends and asking them to get involved. They can write to the White House and also to the prime minister of Denmark by sending an e-mail to him at: stm@stm.dk.

■ I call on people who care about this important issue to write to the prime minister and voice your protest, while informing your friends and networks.

I have to say that some people have written to me and said they are a little afraid to put their name to a protest message, because of the condemnation of any kind of protest now.

What I want to say to you is this: if you give away your right to protest against what you believe is wrong in our world—we're done for—all of us: including the plant and animal kingdoms. The very Earth itself. So get your courage up, if that's what it takes, and move out of that fear zone, thinking not of yourself, but of the whale and dolphin populations, who cannot speak for themselves.

It's up to us now. And, as you have heard said before, we can make a difference.

Yes, and again it is important to write to the White House. I think, at this particular time, that by contacting the Obama administration, which is very pro-environment, we can get the president of the United States to weigh in on this issue.

■ Yes, and it can be a very positive message: "You are a president for change and for ecological advancement. I appeal to you to take active steps to end the mindless slaughter of whales and dolphins, species at great risk in our oceans."

That is right. He can be very influential with the delegation, instructing them how to behave with the International Whaling Commission.

■ Believe it or not, the White House does get those e-mails, and they do read them.

Well, actually they have a computer that reads them, but it picks up keywords. And they know where people stand. The e-mail address is contact@whitehouse.gov.

If you want them to pick up your e-mail, you need words like "whale," "slaughter," and "protest" to appear in your message, so be creative and stay focused on what it is you want the White House to understand from your appeal. Have a sense of urgency about this, because every day counts.

People like Hardy, who has dedicated his life to this quest, and it looks like me now—I'm onboard now—appeal to you. The whales and dolphins need you to survive. We need you, the voice of humankind, to rise to the challenge.

Remember that this issue is not only about the health and welfare of the dolphins and the whales. It is also about the sustainability of the greatest protein source that has ever existed in the world—by that, I mean the fish in the oceans. We have overfished them; we are now poisoning them to the point that, in many fish species, you should not eat any of it. I'd like to take a moment to talk about this.

Absolutely. I am here, talking with the Newfoundlanders about how the fish are disappearing, and I agree it is a very important topic to discuss.

I was in Newfoundland, back in 1984, and did see the humpbacks. At the time, they were talking about the danger to the cod fisheries. That is what originally brought the explorers to North America. It was such an abundant fishery that it fed Europe for several hundreds of years. It has now been, essentially, wiped out. The great cod fishery of the northwest Atlantic has been decimated and, even after they put in controls, it has not recovered. You're really in ground zero there in Newfoundland, with one of the greatest ecological disasters that has ever occurred.

And I am also seeing the recent discovery of oil here—apparently they have discovered huge reserves under the seafloor, here in St. John's Island. So, where cod fisheries used to thrive in pristine bays and some of the purest waters, what I look out on from my hotel window is a number of oil tankers in the harbor. The local people are happy about this—it brings economy to the region. But seeing oil tankers where cod and other fish once thrived breaks my heart.

People have short-term vision when it comes to exploiting resources, and the long-term sustainability of those resources. It seems to almost always play second to the immediate gratification of grabbing what they can from the Earth ... while they can. This, they misguidedly believe, is "prosperity."

Bear in mind too that, while oil is highly profitable for certain people, they can't eat it. You look back not too many years ago and fish was a very inexpensive form of protein. Today it's thirteen to seventeen dollars a pound....

■ *And a large amount of that weight we're paying for is concentrated mercury and other heavy metals and contaminants.*

Yes—the important thing to know is where your fish comes from—and buy accordingly. A few tips: wild cod, Alaskan salmon, and halibut are still very healthy and very sustainable. They have low levels of contaminants in them. Do not, however, eat farmed fish.

■ *Basically, Hardy, you're talking about ninety percent of the fish available to the public! That's what we get in the supermarkets—it's all farmed.*

Well, it is loaded with pesticides and PCBs, and they also use antibiotics.

■ *And do they use growth hormone?*

I'm not sure about that. What they do, actually, is force-feed the fish. They feed them massive amounts of other fish that they catch, so that they grow very fast. That is why farmed fish tend to concentrate the pollutants, because they have eaten so much other fish. Then they put in the antibiotics against various bacteria, because the fish are living in such tight environments.

■ *Yes, I have seen these farms here. There are cages out in the bay waters, not so big—it's a fish zoo for breeding food. The conditions are utterly inhumane. Certainly we can find alternatives to eating this food.*

Yes, so be careful of eating farmed fish, and be very wary of eating large fish—like tuna and swordfish. These species contain such levels of mercury that you can be contaminated by eating very little.

BlueVoice has a kit that tests for mercury, available from our site. You take a snip of your hair and then send it to the University of North Carolina lab, where it is analyzed for mercury. You get a report of how much mercury you have in your body—the overall results of the tests are then put into a data base, so that we are able to get some idea of how vast the problem is.

If you're contemplating becoming pregnant, for example, you should be tested for your mercury levels. If you are pregnant already, you shouldn't touch these big fish products, especially sushi.

Pregnant or not, if you do eat a lot of sushi, or these other big fish, you should do the test.

■ *What are the symptoms of mercury poisoning?*

Well, it varies, of course. It attacks the neurological system: you can feel shaky; you can develop speech impediments; have a sense of fogginess; your hands can shake. They have now discovered that there are a number of other diseases that were never associated with mercury ... and are now believed to be so.

■ *Like Parkinson's?*

Yes, and heart disease. The mercury affects the way the nerves transmit information in our bodies. If the nerves are giving bad information to the heart, you can develop heart disease. The people who came out with this information, ironically, were Danish scientists, working in the Faeroe Islands, which is a protectorate of Denmark. They came out with this information and they told the people there, who eat pilot whales, not to eat them, because it is dangerous to health and causes Parkinson's, heart disease, and other diseases. They also proved that it caused developmental difficulties in the children of the Faeroe Islands.

■ *And yet, not too long ago, some images came around through a number of Internet networks of the slaughter of hundreds upon hundreds of dolphins in the Faeroe Islands. Again, the bay was full of the blood of a massive slaughter of dolphins. Why? It seems to be a rite of passage—some celebration of masculinity for young men there and they do this for fun and a cheap thrill ... and eat the meat too. As brutal as it is to see these things, they do stir you into action.*

To think that a human mind can find pleasure and entertainment in massively slaughtering these beautiful, innocent beings is unacceptable—pure and simple.

These people clearly develop some kind of blood lust, and they clearly have no empathy for the living. This is what baffles me—how anyone can have no empathy for fellow creatures, subjecting them to this suffering and unnecessary and brutal death, for their entertainment.

■ *And what suffering! They slit their throats, while they are still alive, and drag them, quivering in the last throes of life, out of the bay of blood. I don't want to focus on this, but it is reality in these worlds. It is so important that we recognize that this is going on. It's easy to look away, thinking you don't want to give energy to the negativity, but we are responsible for changing this. We are supposed to be the superior species on this planet, although I am sure you will agree, Hardy, that that is debatable.*

Well, we are the self-appointed *Homo sapiens*. We named ourselves that!

■ *Right—what we do know for sure is that we do the most damage out of all the other species. There doesn't seem to be much "sapiens" in that!*

The sperm whale's brain is the biggest brain on the planet ... so we're not necessarily the smartest.

I have had some wonderful, friendly encounters with sperm whales in the Galápagos and down in Dominique, where I have been within eight to ten feet of a massive sperm whale, or in some cases, several of them. I have spent forty-five minutes with a mother and her calf, and looked into her eyes. I have watched tender scenes of one of the young sperm whales, lifting a newborn with his pectoral fins, up to the surface. These animals experience our same emotions— they have thoughts, they have memories, they have a sense of beauty and a sense of love for one another.

■ *And a sense of community ... they travel in families and communities, they are joyous—they are amazing. Nowhere in nature, though, do we see animals killing for the pleasure of it—or for sheer entertainment. That seems to be a strictly human behavior and emotion.*

Just imagine how the world would be, if we didn't hunt and kill whales and dolphins. They already show, in places where they are protected, that they become extremely friendly. What more wondrous thing than to have trust and communication with an animal, like a whale or a dolphin? For me, it has been a miraculous thing to have lived in my life.

■ *Me too. In my limited experience out at sea, I have also had the joy of looking into the eye of a whale. When they make that eye contact with you, it changes your life. It can never be the same after that. It's poignant, it's deep, and you can really feel that higher intelligence coming through.*
It is humbling . . . and it is so, so sacred.

I completely agree. I've been looked at by a lot of sharks and it is not the same look!

■ *Ha! I would not be that excited about eye-to-eye contact with a shark!*

Yes, I was just differentiating there. A shark kind of cruises by you and looks at you and basically says, "I can't eat this," and then cruises on. But when you are with a dolphin or a whale, you see that there is an intelligent being behind that eye. And that being is looking at you, probably thinking: "I wonder what this awkward creature is, here in my realm—it seems to be harmless and it seems to be curious about me, so let's investigate. . . ."

Now, the dolphins in the Bahamas, where we've been for over thirty years—they are just downright friendly. They like to play with us. They bring their calves to us. There have been occasions when they leave their calves with us, at the front of the bow, and go off to feed.

Now, why would they do that?

■ *. . . because they're using you as babysitters!*

They are! They figure a shark is not likely to come up to this very large vessel, so we get in the water and we swim around with the calves, and the mothers go off in the water, maybe grab a little food on the way, and they come back, and their calves are safe. That is a level of trust that is significant, and it is amazing to observe. You can't mistake it for anything else.

I've had a hammerhead shark coming straight at me. There are different species of sharks in the world, and most of them you don't have to worry about, but this one was a really big Atlantic hammerhead shark, and they are dangerous.

This shark was headed straight for me, and I had my camera up and started to roll—there wasn't much else to do. Four dolphins, fairly young male dolphins, came rocketing in, like a squadron of fighter planes, and they dove on the shark, emitting their sonar—very high intensity—they were blasting the sensory perception of the shark. When that happened, that shark tried to get away as fast as he could: he couldn't move out fast enough!

Fortunately, the camera was rolling, so I have it all on video.

It's in a film we did called *The Dolphin Defender*. You just can't argue the fact that these dolphins were interposing themselves between me and the shark. They handled it so easily—it wasn't like they were scared of it. They ganged up on it and chased it out of town.

■ *You had started to talk about the sounds earlier, and I would like to touch on that here. I heard a tape recording of a marine biologist who was determined to communicate with the whales, and who did significant work studying their musical patterns. He found the humpbacks he was studying had a cycle of twenty-five minutes of distinct patterns and sounds in their musical calls. He apparently played a clarinet close to the water level, and a humpback responded to his notes. He played sounds to the whale and the whale was actually trying to imitate those sounds! I found it incredible, listening to a great whale trying to find harmony with the music, and follow the notes.*

Yes, they do respond to that. There is a man named Jim Nolan who has done this all over the world with whales and dolphins. They are predominantly acoustical beings, whereas humans are predominantly visual, so they are naturally going to have an interest in sound. Some sounds appeal to them and others do not. What we found, when we first started playing music to dolphins (back in 1978), is that the dolphins would come from miles around ... but they pretty quickly lost interest. You have to keep something new coming at them, or they will kind of drift away. That is why we came up with this computer system. In the early days, we could track their sounds into our computers and just play that sound back to them. By the time we went out in 2004, we were able to record their sounds and then play them back to them instantaneously. They probably thought, "Hey, these guys are getting smarter!"

■ *Or maybe they were thinking you were becoming better musicians!*

Ah! Dear Hardy, won't it be wonderful to make that breakthrough, where we are able to get through to the whales and dolphins, and finally understand what they are saying and feeling?

We are working on this for alien contact: the SETI program is hoping to intercept a transmission from space and break the code, so why not? Let's break through to our cetaceans, for surely there is a great language and intelligence in our oceans, and these majestic beings are just waiting for us to understand and communicate with them.

I would like to say that the best way to do that is to protect them from direct harm, and from the chemical soups that we inject into the ocean.

■ *And I would like to add, from a metaphysical point of view, that the great whales and the dolphins are weaving the sounds of the ocean—helping to hold the oceans in balance. I am writing about that now and will be releasing that book, Before We Leave You, in 2011. But for the moment, I would like to speak in earthly, pragmatic terms of helping these beings.*

Again, they don't have a voice to protect themselves from the unloving hand of murderous man. We have to do that "alla grande," as the Italians say—big-time. So, voice your protest, write to our leaders, and get involved.

I think that we have reached an inflection point right now. **The time is now.** The people who need to get to work are all of us. The oceans are in really serious danger. People need to inform themselves about it ... and get active.

■ *One way that helps governments and private organizations find a resolution to the economic benefit of killing whales is the boon to tourism for whale watching. If you can, seek out a conscientious, eco-friendly whale watching organization that is taking people out to get close to the whales and to teach people about the life in our oceans. Of course, I'm not talking about the party boats. If you do get close to these gorgeous creatures, it will move you in a way you may never have been moved before.*

They are asking us for help, us humans—the ones "with the brains." They need us to survive ... and we need to show them how they are loved, honored, and revered by the awakened amongst us.

Surely, we can find the time to dedicate to that vision and that quest. Without the great whales and the dolphin beings, the music of the seas dies forever.

■ ■ ■ ■ ■

Find out more about BlueVoice and the work that Hardy Jones and his colleagues are doing, saving dolphins and whales, protecting the oceans, on his Web site: www.bluevoice.org.

Acknowledgments

Many thanks to so many people whose lives are forever dedicated to bringing a new vision of hope and empowerment to a world in need of vision and direction: not only the brilliant minds who have generously shared their deepest thoughts and perspectives in the interviews that make up this work—but all the lightworkers of Planet Earth. You come in with a mission, you walk your talk, and you serve the greater good of all humanity, the environment, the living beings around us—and the planetary deity, Gaia, herself.

My gratitude goes out to Donald Newsom, the director of bbs radio.com, for his tireless commitment to getting the voice of freedom out to the world—despite the forces that would prefer that he, and all of us who work through the station, be silenced. As always, I am so grateful to the dedicated team at North Atlantic Books, continuously providing the support and the platform from which I am given the liberty to bring forward what I perceive as truth. *"Grazie!"* to Patrizia Bertolotti, who encouraged me to create the book from these incredible interviews. Many thanks to Fred Hageneder, for being able to create his cover artwork from the pictures in my mind, and to Jayne Rose, for her work transcribing a portion of the interviews, a tedious job at the best of times.

To all the friends and fellow journeyers, lightworkers everywhere, who connect on so many levels, sharing their love and compassion for our brilliant planet, Gaia—you are the greatest gift of all.

And always, my love to Sara, on high; my beloved soulmate, Franco; Rocky, Tosca, and Kika. You are the caretakers of my soul.